INTERNATIONAL ENVIRONMENTAL GOVERNANCE

Volume 2

Getting to Grips with Green Plans

National-level Experience in Industrial Countries

Full list of titles in the set
International Environmental Governance

Volume 1: Acid Rain in Europe
Volume 2: Getting to Grips with Green Plans
Volume 3: The Global Politics of Pesticides
Volume 4: Acid Earth
Volume 5: Earth Summit II
Volume 6: Fair Weather
Volume 7: Future Generations and International Law
Volume 8: Greening International Institutions
Volume 9: A Guide to EC Environmental Law
Volume 10: Managing the Planet
Volume 11: Protecting the Atmosphere
Volume 12: From Earth Summit to Local Agenda 21
Volume 13: Futile Progress
Volume 14: Yearbook of International Cooperation on Environment and Development 1998–99
Volume 15: Yearbook of International Cooperation on Environment and Development 1999–2000
Volume 16: Yearbook of International Cooperation on Environment and Development 2001–02
Volume 17: Yearbook of International Cooperation on Environment and Development 2002–03
Volume 18: Yearbook of International Cooperation on Environment and Development 2003–04

Getting to Grips with Green Plans

National-level Experience in Industrial Countries

Barry Dalal-Clayton

earthscan
from Routledge

Routledge
Taylor & Francis Group
LONDON AND NEW YORK

First published 1996 by Earthscan

Publisher 2022 by Routledge
4 Park Square, Milton Park, Abingdon, Oxon OX14 4RN
605 Third Avenue, New York, NY 10017

Routledge is an imprint of the Taylor & Francis Group, an informa business

Publisher's Note

The publisher has gone to great lengths to ensure the quality of this reprint but points out that some imperfections in the original copies may be apparent.

ISBN 978-1-844-07986-5 (hbk) (Volume 2)
ISBN 978-1-84407-984-1 (International Environmental Governance set)
ISBN 978-1-84407-930-8 (Earthscan Library Collection)

A catalogue record for this book is available from the British Library

Library of Congress Cataloging-in-Publication Data has been applied for

GETTING TO GRIPS WITH GREEN PLANS

Barry Dalal-Clayton

with contributions from Izabella Koziell, Nick Robins and Barry Sadler

GETTING TO GRIPS WITH GREEN PLANS

National-Level Experience in Industrial Countries

EARTHSCAN

Earthscan Publications Ltd, London

First published in 1996 by
Earthscan Publications Limited
120 Pentonville Road, London N1 9JN
Tel: 0171 278 0433
Fax: 0171 278 1142
E-Mail: earthinfo@earthscan.co.uk

A catalogue record for this book is available from the British Library

ISBN: 1 85383 428 9 (Paperback)

Copy-edited and typeset by Selro Publishing Services, Oxford

Earthscan Publications Limited is an editorially independent subsidiary of Kogan Page Limited and publishes in association with the International Institute for Environment and Development and the WWF–UK.

Contents

Boxes, Tables and Figures viii
Acknowledgements xi
Preface xiii
Executive Summary xv

PART I: OVERVIEW AND SYNTHESIS

1. **Introduction** 3
   Background 3
   The Research 6

2. **Approaches and Stimuli** 9
   The Stimulus of UNCED and Environmental Concern 9

3. **Focus of Green Plans and Strategies** 17
   Sustainable Development Strategies 18
   Environmental Strategies and Plans 21
   Special Focus Studies 23

4. **Duration, Time Frames, Mandates and Management Approaches** 24
   Duration of Development Process 24
   Mandates and Terms of Reference 24
   Strategy Management Processes 27

5. **Participation in Green Plans and Strategies** 29
   Participatory Approaches 30
   Consultative Approaches 32
   Predominantly Internal Government Processes 40

6. **Links to Other Strategy and Planning Processes** 43
   Regional Strategies 43
   Links to Budget Processes 50
   Links to National Planning 51
   Sub-National Strategies 53
   Local Agenda 21s 55

Convention Strategies 59

7. **Politics, Governance and Legislation** *61*
Political Influences 61
Cabinet and Parliamentary Review/Debate 64
Legislative and Institutional Consequences of Strategies 64

8. **Conclusions** *67*
Towards Sustainable Development Strategies 67
Developed and Developing Country Approaches Briefly Compared 68
Challenges 71

PART II: COUNTRY CASE STUDIES

9. **Australia** *75*
The National Strategy for Ecologically Sustainable Development (NSESD) 77

10. **Canada** *90*
The Green Plan 92
The Projet de société 100

11. **Denmark** *109*
Foundations for Strategic Environmental Planning 109
Nature and Environment Policy 110

12. **France** *115*
Plan National pour l'Environnement 115
French Commission for Sustainable Development 123

13. **Latvia** *126*

14. **The Netherlands** *133*
National Environmental Policy Plans 133
Action Plan: Sustainable Netherlands 144

15. **New Zealand** *148*
The Resource Management Act and the Environment 2010 Strategy 148

16. **Norway** *165*
Reports to Parliament on the Follow-Up to the Report of the Brundtland Commission (No 46), and on UNCED (No 13) 165

17. **Poland** *174*
National Environmental Policy 174

18. **Sweden** *182*
Towards Sustainable Development in Sweden: Government Bill 1993/4:111 182
An Environmentally Adapted Society: The Action Programme of the Environmental Protection Agency: Enviro '93 186

19. **United Kingdom** *191*
Sustainable Development: The UK Strategy 191

20. **United States of America** *205*
    President's Council on Sustainable Development *206*
    National Environmental Goals Project *220*

21. **The European Union** *223*

22. **Green Planning and Strategy Processes in Some Other Industrial Countries** *236*
    Austria *236*
    Germany *237*
    Ireland *238*
    Japan *239*
    Portugal *241*
    Russia *242*

Appendices *244*
1. Questionnaire for Structured Interviews *244*
2. Useful Contacts *247*

Acronym *253*
References *257*
Further Reading *266*
Index *269*

Boxes, Tables and Figures

Boxes

1.1 The International Network of Green Planners *4*
1.2 Key Lessons and Guiding Principles for National Sustainable Development Strategies *6*
3.1 Environmental Space and Ecological Footprints *20*
5.1 A Typology of Participation in Policy Processes and Planning *30*
5.2 Canada's Projet de société *33*
5.3 The Round-Table Experience in Canada *34*
5.4 Why Consultation Became a Late Feature of the Canadian Green Plan *37*
6.1 Recent European Environmental Initiatives *46*
6.2 The Environmental Action Programme for Central and Eastern Europe, and Environmental Programme for Europe *48*
6.3 Objectives of the North American Agreement on Environmental Cooperation *50*
6.4 Norway's Green Budget *52*
6.5 New Zealand's Resource Management Act (1991) and Planning *54*
6.6 Examples of Sub-National Strategies in Canada *55*
6.7 Examples of 'Independent' Local Strategy Initiatives in Canada *59*
8.1 Key Tasks in the Strategy Process in Developing Countries *69*
9.1 Main Elements of Strategy Processes in Australia *76*
9.2 Goal, Core Objectives and Guiding Principles of the Australian National Strategy for Ecologically Sustainable Development (NSESD) *79*
9.3 Australian National Strategy for Ecologically Sustainable Development (NSESD) — Main Contents *80*
9.4 The NSESD Process *81*
9.5 A View on the Need to Legislate for Sustainable Development in Australia *84*
10.1 Canada's Guide to Green Government *99*
10.2 Choicework *102*
10.3 Sustainability Goals for Canada *104*
10.4 SustainABILITY *105*
11.1 The Scope of Strategic Environmental Planning in Denmark *111*
12.1 Principles for Action in the PNE *117*

12.2 Summary of the PNE *118*
13.1 NEPPL Goals and Priority Problems *128*
14.1 Key Instruments in the Implementation of the NEPP, 1989 *136*
14.2 Aims of NEPP2, 1993 *137*
14.3 Covenants in the Netherlands *141*
14.4 Environmental Space *145*
14.5 Action Plan: Sustainable Netherlands *146*
15.1 Some Economic Reforms and Related Issues in New Zealand *150*
15.2 Maori Claims *152*
15.3 Ecological Characteristics of New Zealand *153*
15.4 Vision of New Zealand's Environment 2010 Strategy *154*
15.5 Principles for Integrating Environment, Society and Economy *154*
15.6 *Environment 2010 Strategy* — Environmental Management Agenda *155*
15.7 Priority Issues and Environmental Goals in New Zealand *156*
15.8 New Zealand's Sustainable Land Management Strategy *158*
15.9 New Zealand's Sustainable Water Management Strategy *159*
16.1 Strategy Processes in Norway *166*
16.2 Economic Considerations in Norway's Report No 46:1988–9 to the Storting *167*
18.1 Summary of Swedish Government Bill 1993/4:111 *184*
18.2 Scope of Sweden's Enviro '93 Programme *188*
19.1 Examples of Local Agenda 21 Initiatives in the UK *197*
19.2 North-South Linkages and Local Agenda 21 in the UK *200*
20.1 Executive Order No 12852 *207*
20.2 US National Goals Towards Sustainable Development and Example Indicators of Progress *210*
20.3 Recommended Actions to Create a Community-Driven Planning Process in the USA *215*
20.4 Some State Initiatives Towards Sustainable Development *217*
20.5 Draft Environmental Goals for Clean Air in the USA *222*
21.1 Sustainable Europe Project *234*

Tables

1 Basic Comparisons Between Developed and Developing Country Strategy Processes *xx*
1.1 Study Interviews: Countries and Strategies *7*
2.1 Strategies Characteristics Compared *10*
2.2 Key Characteristics of NEAPs in Central and Eastern Europe *15*
4.1 Duration of Strategy Preparation *25*
4.2 Strategy Time Frames *26*
8.1 Comparisons Between Developed and Developing Country Strategy Processes *71*
10.1 Example of Choicework Table *103*
21.1 European Objectives, Targets and Actions for Climate Change *225*
21.2 European Municipal Waste Management Objectives, Targets and Actions *226*

21.3 Extract of Comparison of Agenda 21 and the European Fifth Environmental Action Programme *230*

Figures

15.1 Process of Setting Priorities and Monitoring Outcomes for New Zealand's Environment 2010 Strategy *163*
20.1 President's Council on Sustainable Development: Key Steps to the Submission of the National Sustainable Development Strategy *213*

Acknowledgements

The research on which this book is based was undertaken with financial support provided to IIED by the Royal Norwegian Ministry of Foreign Affairs (MFA) under an MFA–IIED trust agreement. Particularly, thanks are due to Mrs Randi K Bendiksen and Dr Eirek Jansen of MFA for their interest and support for this work.

The book is based substantially on a comparative analysis of information provided during semi-structured interviews and subsequently by individuals centrally involved, in their countries, in the management of national 'green planning' processes. Without their help and cooperation, this work would not have been possible. Special thanks are due, in alphabetical order by country, to:

Paul Garrett, Laurie Hodgeman, John Scanlon and Andy Turner (Australia); Stephen Blight, Ron Doering, Sandy Scott, Dana Silk, Richard Smith and Wilma Vreeswijk (Canada); Jytte W Keldborg and Marianne Rønnebæk (Denmark); Robert Donkers and Robert Hull (European Commission); Pierre Guelman, Michel Hors and Jacques Theys (France); Zigfrids Bruvers, Ilona Lodzina and Valts Vilnitis (Latvia); Maria Buitenkamp, Gerard Keijzers and Annette van Schreven (The Netherlands); John Gilbert, Lindsay Gow, Kevin Steel and Bob Zuur (New Zealand); Paul Hofseth (Norway); Glen Anderson, Marek Haliniak, Andzej Kassenberg, Stefan Kozlowski, Agata Miazga and Tomasz Zylicz (Poland); Jon Kahn, Mats Olson, Sture Persson and Inger Vilborg (Sweden); John Stevens (UK); Derry Allen, Julie Frieder, Gary Larsen, Jonathan Lash, Jan McAlpine, Molly Harriss Olson and Donna Wise (USA).

Grateful thanks are due to Izabella Koziell (IIED) for undertaking the interviews in Latvia and Poland, Nick Robins (IIED) for those at the European Commission, and Barry Sadler for help with those in Australia and New Zealand. Information on the strategy processes in Ireland and Russia was kindly provided by Geraldine Tallon and Renat Perelet, respectively.

Helpful comments on the manuscript have also been provided by IIED colleagues, particularly Koy Thomson and Steve Bass, while Nick Robins contributed Chapter 21 and text on the situation in the European Union in Chapter 6.

The analysis and structure of this book builds on the experience of IIED working with many other institutions, organizations and individuals in assisting or promoting the development of strategies for sustainable development or their near equivalents. Particularly in designing the questionnaire, which formed the basis for the structured interviews, the author has drawn from an IIED issues paper, 'National Sustainable

Development Strategies: Experience and Dilemmas' (Dalal-Clayton et al, 1994); and also from work undertaken by IIED and IUCN during 1992–4.

Finally, the views expressed in this book, except where otherwise stated, are those of the author and do not represent an official view, either of IIED or any of those persons interviewed.

Preface

Developed countries have a long history of planning. Over the last few years, many have undertaken or embarked on national planning exercises to deal with growing environmental problems — green plans. Following UNCED, many countries are attempting to address the issue of sustainable development and are considering how to respond to Agenda 21 at a national level. Agenda 21 calls for countries to develop national sustainable development strategies (NSDSs).

In approaching this challenge, it is fair to say that the North has much to learn from the experience of the South, and vice versa. Over the last 15 years, much experience has been gained through developing national conservation strategies, National environmental action plans, tropical forestry action plans and many similar approaches. Many of these initiatives in developing countries have been funded, assisted or promoted by bilateral development agencies, UN organizations and multilateral development banks. Efforts in developed countries have been funded domestically, mainly by governments.

Since UNCED, considerable effort has been made to review and distil this past experience and to draw out lessons that can help guide processes to develop and implement NSDSs or their equivalents. IIED and IUCN have been working both collaboratively and independently on tracking and analysing such strategy experiences in many countries around the world, through regional workshops and case studies. This work resulted in the joint publication of *Strategies for National Sustainable Development: A Handbook for their Planning and Implementation* (Carew-Reid et al, 1994), regional compendia of case studies (IUCN, 1993a and 1993b); and, subsequently, a series of IIED issues papers on various aspects of NSDSs (Dalal-Clayton et al, 1994; Bass, Dalal-Clayton and Pretty, 1995; Bass and Dalal-Clayton, 1995). Various other organizations have also conducted workshops and reviews, mainly of developing country experience, including the Network for Environment and Sustainable Development in Africa (NESDA), OECD, UNDP, and the World Bank.

Developing countries are asking increasingly what experience the North has had, and what lessons they can draw from this. The International Network of Green Planners was established in 1992 and will hopefully provide a platform for the continued sharing of experience between strategy practitioners and green planners from the North and South.

The problems faced by developing and developed countries in preparing NSDSs are usually quite different. Most developing countries are occupied with achieving economic development, through industrialization where this is possible, and by expanding production. By comparison, one of the key issues for sustainable development in most developed countries is dealing with the problems caused by high levels of consumption, by existing industries and by technology-based economies (for example, pollution and waste).

This book examines the recent experiences, including successes and difficulties, of industrialized countries in developing their national strategies for sustainable development, green plans or near equivalents. Part I (Chapters 1 to 8) provides an overview and synthesis of the main approaches and processes adopted, while Part II (Chapters 9 to 21) presents details of 20 green planning initiatives in ten selected countries and in the European Union.

It is hoped that this study will be of interest to policy-makers, planners, and strategy practitioners in both developed and developing countries.

Barry Dalal-Clayton
10 July 1966
IIED, London

Executive Summary

This study reviews and compares 20 recent green planning initiatives in 12 industrialized countries: Australia, Canada, Denmark, France, Latvia, The Netherlands, New Zealand, Norway, Poland, Sweden, the UK and the USA, together with regional initiatives in eastern and western Europe (which are listed in Table 1.1). The majority are government-sponsored initiatives, but two were conducted by non-governmental organizations, and one was a programme of the European Union. They include activities that preceded the 1992 UN Conference of Environment and Development (UNCED) and others that were undertaken in response to UNCED, particularly Agenda 21. The initiatives include a wide range of different approaches (environmental plans, strategies, legislative instruments, reports to parliaments, and sustainable development commissions).

The book is presented in two parts. Part I (Chapters 1–8) provides an overview and synthesis of the main green planning approaches and processes followed in the countries studied. Part II (Chapters 9–21) presents details of the green planning initiatives in each of the selected countries and in the European Union.

Part I

Chapter 1 provides a background to the study, reviewing the origins and scope of 'green planning', and the challenges of developing a national sustainable development strategy (NSDS) — as called for in Agenda 21. A number of dilemmas likely to face those charged with developing an NSDS are outlined, based on lessons from experiences mainly in developing countries. Drawing from these perceived dilemmas and lessons, a questionnaire was designed (Appendix 1) and used for a series of semi-structured interviews with key individuals, who were involved in the selected plans and strategies and have been responsible for managing those processes. These interviews provided the basis for the country case studies presented in Chapters 9–21. The aim was to focus on the particular perspectives of these key players.

In *Chapter 2*, a detailed comparison is provided of the key characteristics of the different approaches in the strategies and plans studied. As already noted, some initiatives preceded UNCED; others were a response to it. Many were undertaken as a direct response to public concerns about the environment. Most had an official government mandate. The involvement of cabinets and parliaments is discussed. Some led to legal and/or institutional changes. The focus of the different plans and strategies vary. For example, some have a dominantly environmental focus, some are concerned with the broader issues of sustainable development, a number are concerned mainly with federal areas of responsibility, some have set targets and time horizons while others do not, and a few are designed around special issue studies. Most of the initiatives are predominantly internal government exercises — only a few are independent. They mainly involve cross-government and inter-departmental processes, but some have been undertaken exclusively within single ministries/agencies. Extensive stakeholder participation has been a feature of only a few of the strategies and green plans concerned, but most have involved some form of consultation with industry, NGOs and the public. The stimulus for green plans and strategies provided by UNCED, and the demands for government action in response to growing public concern about the environment in many countries, are discussed.

Chapter 3 provides a more detailed consideration of the focus of green plans and strategies reviewed in three categories — strategies concerned predominantly with sustainable development; environmental strategies or plans; and special focus studies. However, while these distinctions can usefully be made, in reality there is a continuum of approaches, and experience has shown that, with time, environmental plans and strategies can evolve to become sustainable development strategies.

In trying to deal with sustainable development, a range of different approaches are evident. For example, in Australia, the concept of ecologically sustainable development has been the main focus. The Canadian Projet de société devised innovative choicework tables to address trade-off issues. In the USA, the President's Council on Sustainable Development has developed principles for sustainable development that aim to integrate environmental, social and economic goals and objectives. The Dutch Friends of the Earth pioneered the concept of 'environmental space', while the UK Strategy cautiously raises the concept of 'ecological footprints' (but without resolving the government's view).

Environmental strategies have generated a broad range of responses. For example, the Canadian *Green Plan* (1990–6) (Government of Canada, 1990a) — arguably the 'mother' of green planning — was primarily concerned with environmental decision-making and was an action plan to address specific issues (for example, climate and fisheries). The various Dutch national environmental policy plans NEPP (VROM, 1989), NEPP+ (VROM, 1990), and NEPP2 (VROM, 1993a) were concerned with a range of environmental source and process themes (for example, acidification, waste disposal) and set environmental targets. The environmental goals projects of the USA's EPA also set goals around environmental targets. Norway introduced the idea of a 'green budget'. Some countries (for example, Denmark) have instituted periodic environmental progress reports. The French *Plan national pour l'Environnement* (French MOE, 1990a) provided a vehicle for the reform of the public administration of environmental management, while Poland's *National Environmental Policy* (PolMEP, 1990a) aimed at 'green reconstruction' of particular economic sectors. Sweden's Enviro '93 programme also aims to shift environmental responsibilities to sectors.

In *Chapter 4* the duration, time frames, mandates and management approaches of green plan and strategy processes are examined. They have varied between about six months and three years and most have adopted or been based on some time frame for the visions they contain or for implementation of actions. Usually, strategies initiated by governments have had some form of official mandate or terms of reference issued by the head of government or a minister, or drawn up by civil servants and subsequently endorsed by the government.

Few strategy documents describe the process by which they were developed, though Australia's *National Strategy for Ecologically Sustainable Development* (COA, 1992a) is a notable exception. Governments have established a wide range of mechanisms with which to develop and manage the green plan and to review strategy processes. These include core teams; steering committees; cross-government negotiations; representative councils and fora involving senior figures from industry, as well as academics and NGOs; advisory groups and round tables; informal meetings; working papers released for public comment; seminars, workshops and public meetings; drafting teams; Cabinet scrutiny; and, in a few cases, parliamentary approval.

Non-governmental or independent strategies tend to adopt additional and more innovative approaches, including a secretariat performing a 'facilitating' role rather than coordinating/directing affairs; participatory stakeholder round tables and assemblies deciding directions and taking decisions; and inputs by volunteers.

In *Chapter 5*, the issue of participation is dealt with in detail. The more open green plans and strategies have experimented with different approaches to participation, including stakeholder round tables, providing financial support to enable NGOs to become involved, funding NGOs to undertake commissioned work, involving target groups, and adopting traditional approaches (for example, in New Zealand traditional *hui* were used for meetings with Maori organizations). The Canadian Projet de société arguably represents the most participative national-level strategy process so far attempted. It was developed as a consensus process involving over 80 businesses and government and independent organizations in a National Stakeholders' Assembly. The round table approach, pioneered by Canada, is reviewed.

But most green planning and strategy processes in industrial countries have adopted a 'consultative' approach in which participants are restricted to listening and providing information (through, for example, public inquiries, media activities and telephone 'hot-lines') or are consulted (for example, through working groups and meetings held to discuss plans and policies). They have little effective say in building a consensus around the main elements of the strategy, or in the decision-making about either the policy or the strategy and its various components. In the Netherlands, the NEPP2 process placed great emphasis on persuading target groups to 'participate' in discussing what changes they should make and to become involved in monitoring implementation. This led to the negotiation between industry and government of innovative covenants to help meet NEPP targets and complement existing legislation.

A number of plans and strategies have been developed mainly as internal government processes: examples in Western Europe are Denmark's *Nature and Environment Policy* (DanMoE, 1995) and Sweden's Bill 1993/94.111, *Towards Sustainable Development in Sweden* (SwedMoE, 1994); and, in Eastern Europe, Poland's *National Environmental Policy* (PolMEP, 1990a). It needs to be added, however, that after decades of central planning, central and eastern European governments are not oriented towards people's participation, and the people there are not accustomed to 'participating' in government decision-making.

The links between national strategies and green plans on the one hand, and between other strategy and planning processes on the other hand, for example regional and convention-related strategies, are discussed in *Chapter 6*. Particular attention is paid to the efforts launched in 1993 by the European Union to develop a strategy for sustainable development — the Fifth Environmental Action Plan, building on a series of previous five-yearly environmental action plans. However, there has been no attempt to coordinate national planning and strategy exercises. In North America, Canada's Green Plan interfaced with various regional initiatives, for example, the Circumpolar Conservation Strategy and the North American Free Trade Agreement.

In central and eastern Europe, action to halt and reverse environmental degradation is being promoted and coordinated through a regional Environmental Action Programme (EAP). The EAP document (UNECE, 1995) has been used in some countries in the region as a 'handbook' for the development of national environmental plans.

Some strategies have been directly linked to government budgetary processes (for example, Canada's *Green Plan*, the Dutch NEPP2, New Zealand's *Environment 2010 Strategy*, and the *Norwegian Report No 46 to the Storting*). It is not easy to determine, from documentation and discussions, the extent to which green planning initiatives and sustainable development strategy processes have seriously influenced, or are linked to, mainstream national planning. However, some links are clear in a few cases – Denmark's 1995 *Nature and Environment Policy*, the 1990 French *Plan national pour l'Environnement*, and New Zealand's 1991 *Resource Management Act*.

Amongst the national plans and strategies studied, only a few were formally linked with (ie built directly on or leading directly to the development of) sub-national strategies at the provincial, territorial or state level. Nevertheless, such sub-national strategies are common, notably in federal countries such as Australia, Canada and the USA. Examples of these are discussed. Consideration is also given to the explosion of Local Agenda 21s and similar initiatives throughout industrial countries — both those linked to national strategies and those initiated independently. In most countries, green plans and sustainable development strategies have been developed independently of national plans required by the conventions on climate change and biodiversity.

In *Chapter 7*, the domestic political influences that have shaped the development of green plans and strategies are examined. The Canadian *Green Plan* (Government of Canada, 1990a) and follow-up initiatives were greatly influenced by the agenda of the Conservative government in the late 1980s, and the programme of the subsequent Liberal government. The fortunes of Australia's 1992 *National Strategy for Ecologically Sustainable Development* (COA, 1992a) were strongly influenced by prime ministerial change. Various other influences are discussed. The involvement of cabinets and parliaments in initiating and/or approving plans and strategies is reviewed, and the legislative and institutional consequences of various strategies is also described.

In *Chapter 8*, some conclusions are presented. The extent to which the green plans and strategies reviewed match up to the requirements of genuine and effective sustainable development strategies is considered.

The government-led processes have been fashioned mainly by prevailing political, bureaucratic and cultural circumstances in the industrial countries concerned, and have usually adopted approaches consistent with routine government practices for such initiatives. Furthermore, the different plans and strategies have been developed to address particular domestic environmental, social and economic conditions and circumstances, which differ in each country. While the initiatives covered can all be described — and indeed are promoted by their principal architects — as green planning processes, in practice they represent a range of quite different approaches (for example, environmental plans, strategies, legislative instruments, reports to parliament, Commission processes) and are aimed at fulfilling a variety of different objectives (some visioning, some goal-setting, some for implementation). They are not equivalent processes and it is not possible to compare them as if they were.

In approaching the challenge of developing national sustainable development strategies, it is fair to say that the countries of the North and the South have much to learn from the experience of each other. Some comparisons are therefore made between approaches in developed and developing countries (see Table 1).

Past research and analysis has shown that a number of steps appear to be common to the more successful strategies in developing countries (see Box 8.1). But this is not surprising given that many of the approaches have followed a basic framework, which has been developed for national conservation strategies and, as experience has grown, has subsequently built on and improved national environmental action plans, tropical forestry action plans and similar initiatives. Furthermore, these approaches have been promoted in developing countries mainly by donors who have provided the financial support and technical assistance to replicate the models in different countries as a framework for aid support. In many cases, the expatriate technical experts and advisers have worked on strategies in several countries and have translocated their experience and approaches.

But the situation in developed countries is entirely different. No common approach is apparent in the processes adopted. As already noted, they have all been fashioned according to domestic agendas and have followed national government styles and cultures rather than those of external agencies. It is still too early to say whether any of the basic requirements that appear to characterize strategies in developing countries apply to those in developed countries (see Box 8.1).

For example, it is logical that green plans and strategies in industrial countries should move closer towards the 'ideal' of sustainable development strategies if they are cyclical, i e if they are periodically revised to take into account feedback and lessons from review following implementation, and thus to become genuine 'learning by doing' processes. But to date, of the initiatives reviewed in this study, the only genuine second-generation processes are the second Dutch *National Environmental Policy Plan*, NEPP2 (VROM, 1993a), building on the NEPP (VROM, 1989) (it is assumed that the NEPP3, planned for 1997, will build further on this experience), and the European Union's Fifth Environmental Action Programme building on previous programmes.

A serious question, which will need to be addressed by industrial countries if they are to make progress towards meeting the challenges of sustainable development, is to what extent will it be necessary to adopt the approaches found to be successful in developing countries (i e as suggested by the key tasks listed in Box 8.1)? In particular, to what extent will it be necessary to move towards being more participative, integrative, and cyclical?

Some of the initiatives discussed in this study have made impressive progress in this direction; others have been little more than environmental planning and policy-making as usual. The question to be asked is whether governments are serious about moving their societies and economies towards a sustainable future, or are merely paying 'lip service' and responding to the issue in a traditional way by driving Agenda 21 into an 'environmental rut'?

***Table 1** Basic Comparisons Between Developed and Developing Country Strategy Processes*

Developed Countries	Developing Countries
Approach	*Approach*
Internally-generated	External impetus (IUCN, World Bank)
Internally-funded	Donor-funded
Indigenous expertise	Expatriate expertise frequently involved
Political action	Bureaucratic/technocratic action
Brokerage approach	Project approach
Aims	*Aims*
Changing production/consumption patterns	Increase production/consumption
Response to 'brown' issues (pollution)	Response to 'green' issues/ rural development
Environment focus	Development focus
Means	*Means*
Institutional re-orientation/integration	Creation of new institutions
Production of guidelines and local targets	Development of project 'shopping lists'
Cost-saving approaches	Aid-generating approaches
Links to Local Agenda 21 initiatives	Few local links
Awareness-raising	Awareness-raising

Another question, which only will be answered in the future, is whether or not national sustainable development strategies will have any lasting influence on the development and implementation of public policy and economic development, and on social attitudes and behaviour. Will they, as some observers predict, merely 'sit on the bureaucratic bookshelves gathering dust like earlier generations of master plans' (Rowley, 1993)? In this report are described examples of approaches and innovations which provide a positive basis for hope.

Part 2

Chapters 9 to 21 represent case studies, each providing details of the key green planning or strategy processes in the selected countries, based on structured interviews undertaken with the coordinators and key individuals involved in these exercises. In *Chapter* 22, brief descriptions are provided of recent initiatives in Austria, Germany, Ireland, Japan, Portugal and Russia.

Finally, two appendices are included. Appendix 1 is the questionnaire used as a basis for the interviews, while Appendix 2 gives details of these individuals and useful contact addresses for those requiring further information on the planning and strategy processes covered in this book.

PART I
OVERVIEW AND SYNTHESIS

Chapter 1 INTRODUCTION

'Green planning' is a term originally applied to plans developed, mainly in industrial countries, to address escalating environmental problems. But some developing countries have also used this terminology (for example, Namibia). It appears to have been first introduced formally in 1989 when, in response to public and political pressure to deal with mounting concern about the environment, the Canadian environment ministry — Environment Canada — embarked on preparing a *Green Plan* for Canada (Government of Canada, 1990a). More recently, 'green planning' has been used in a wider context — as a 'shorthand' description, as Sadler (1996) puts it — to embrace a range of initiatives, including those plans and strategies concerned with broader issues of sustainable development, particularly by members of the International Network of Green Planners, established in 1993 to foster information exchange and learning among strategy practitioners and policy-makers (Box 1.1). However, as we will show in this book, most green plans produced in the industrial countries remain focused on environmental issues; very few (mainly those undertaken independently of governments) have yet attempted to balance environmental, social and economic concerns — a central requirement of moving towards sustainable development.

Background

National sustainable development strategies (NSDSs) are now widely seen as one of the main mechanisms for setting out national approaches to Agenda 21 — the action programme for sustainable development adopted at the 1992 United Nations Conference on Environment and Development (UNCED) — particularly to meet objective 8.3, namely 'improving or restructuring the decision-making process so that consideration of socioeconomic and environmental issues is fully integrated and a broader range of public participation assured' (UNCED, 1992). Agenda 21 recommends that each nation:

> *should adopt a national strategy for sustainable development based on, inter alia, the implementation of decisions taken at the [UNCED] Conference, particularly in respect of Agenda 21. This strategy should build upon and harmonize the various sectoral economic, social and environmental policies and plans that are operating in the country. The experience gained through existing planning exercises such as national reports for the Conference, national conservation strategies and environment*

Box 1.1 THE INTERNATIONAL NETWORK OF GREEN PLANNERS

The INGP was founded in December 1992 by a small group of experts from governments and agencies engaged in strategic environmental management, which met in Washington DC. It is now coordinated by a secretariat housed in the Directorate for Strategic Planning of the Dutch Ministry for the Environment. The secretariat is responsible for registering members (now numbering some 200 practitioners from around the world), facilitating contacts and exchanging information on issues of interest to members, setting agendas within the network, and assisting in organizing international and regional network meetings. INGP communication is assisted by Green Page *— a periodic network newsletter. Another useful tool is the* Green Planners' Guide, *which provides brief professional profiles of INGP members.*

Two international meetings have been held to date. The inaugural meeting in Maastricht, in April 1994, provided a forum for exchange of ideas and experiences and considered policy processes and mechanisms. The second in San Francisco in June 1995 concentrated on issues related to implementing green plans: barriers, stakeholder involvement, problem definition, goal development, and measuring progress, with water management as a focal theme. Both meetings were supported by issue papers prepared in advance by consultants to facilitate discussion.

> *action plans should be fully used and incorporated into a country-driven sustainable development strategy. Its goals should be to ensure socially responsible economic development while protecting the resource base and the environment for the benefit of future generations. It should be developed through the widest possible participation. It should be a thorough assessment of the current situation and initiatives.*
>
> (Chapter 8, Agenda 21, UNCED, 1992)

However, neither Agenda 21 nor the UNCED secretariat provided any guidelines on how to prepare NSDSs, and none have yet been forthcoming from the UN Commission for Sustainable Development. Nevertheless, a number of nations (for example, China and the UK) have already responded directly to Agenda 21 by preparing a strategy, national Agenda 21 or equivalent action plan. Many other countries have now engaged in some form of similar process.

Many developed and developing countries had already prepared, or were well advanced with, some form of environmental plan or equivalent strategy process before UNCED (for example, Canada, the Netherlands, the United Kingdom, Botswana and Pakistan). During the last 15 years, many countries have developed various forms of comprehensive national strategies and plans, which aim, to a greater or lesser extent, to integrate environmental and developmental objectives. These include conservation strategies, environmental action plans, green plans and forestry action plans. From this wide body of experience, many lessons can be drawn on how to approach the process of undertaking an NSDS, as advocated by Agenda 21. In a recent two-year study, IIED and IUCN examined the experiences of over 100 countries in developing and implementing various forms of strategy, leading to the publication of a 'Handbook on Strategy Preparation and Implementation' (Carew-Reid et al, 1994).

There have been several other reviews of experience of strategies and 'green planning' (for example, Hill, 1992; Dalal-Clayton et al, 1994; ERM, 1994a,b,c and d; Bass and Dalal-Clayton, 1995; OECD, 1995a) and case studies of strategies undertaken in different regions (IUCN, 1993a and 1993b). The World Bank has also reviewed its experience in promoting national environmental action plans (World Bank, 1995), while the Regional Environmental Centre for Central and Eastern Europe has recently assessed progress on the development and implementation of National Environmental Action Programmes in CEE countries (REC, 1995).

A number of dilemmas are likely to face those charged with developing an NSDS (Dalal-Clayton et al, 1994): Some of these include:

- *The political context* (for example, structural constraints and inequalities in national and local power structures) will greatly influence how a strategy can be developed and implemented, how values can be defined and/or expressed, and how choices can be made. Some issues that a strategy may have to address are likely to be highly politically charged, for example, land ownership, environmental degradation and poverty.
- *Setting the objective(s)*: different groups are likely to want to achieve different things by preparing a strategy. For a strategy to be effective, the constituency needs to agree on the objectives, and these should determine the process, not the reverse of this (as has often happened).
- *Building strategic capacity*: sustainable development needs an interdisciplinary approach that aims to integrate environmental, social and economic objectives. Achieving such a balanced approach (rather than concentrating on just one dimension) is a major challenge. Past experience suggests that the capacity of agencies, communities and other groups to think and work strategically is at least as important as the strategy exercise or plan itself.
- *Establishing the scope of a strategy*: finding a balance between local, national and international issues, between national and regional strategies and between national and local strategies (for example, Local Agenda 21s); determining which of these approaches should take precedence; and considering how to tackle controversial and uncertain boundary issues (for example, 'ecological footprints' and 'environmental space').
- *Multiple national strategies*: reconciling conflicts, confusion, overlaps (in time, scope and content) when more than one national strategy is undertaken.
- *Consultation versus participation*: who are the 'stakeholders' in a strategy process? Who should be involved and when? What should they be involved with? Who should be targeted? Balancing the benefits and risks of public participation — this issue is perhaps one of the most difficult challenges.
- *Choosing approaches and methodologies*: whether and how to adopt and promote new ways of thinking and practice to accelerate participation, community-self-reliance, and institutional change to handle holistic concepts and uncertainties that an NSDS may require.

The recent work of IIED and IUCN on strategies drew from numerous case studies and from experience discussed at several regional workshops (Carew-Reid et al, 1994). It has enabled a number of *key lessons and guiding principles* to be identified (Box 1.2).

Box 1.2 KEY LESSONS AND GUIDING PRINCIPLES FOR NATIONAL SUSTAINABLE DEVELOPMENT STRATEGIES

- *National sustainable development strategies are cyclical processes of planning and action in which the emphasis is on managing progress towards sustainability goals rather than producing a 'plan' or end product.*
- *They must be genuinely multi-sectoral and integrative, aimed at engaging relevant interests and overcoming institutional and policy fragmentation.*
- *It is crucial to focus on priority issues, and identify key objectives, targets and means of dealing with them.*
- *'Widest possible participation' means sharing responsibility and building partnerships among all concerned — business, community and interest groups, as well as governments — but only where the partners feel it is appropriate.*
- *The approach taken must be adaptive and flexible, recognizing that problems are characterized by complexity and uncertainty, and policy responses and technological capability change over time.*
- *Monitoring, evaluation and learning from experience are keys to a successful strategy, and must be an integral part of the process.*
- *The preparation of an NSDS is an exercise in capacity-building, and should be organized to enhance institutional arrangements, sharpen concepts and tools, foster professional skills and competence, and improve public awareness.*

(Dalal-Clayton et al, 1994)

The Research

This book considers how industrial countries have addressed the dilemmas listed above, examines other difficulties they have faced, and discusses the principles identified in Box 1.2 in relation to their experiences. It focuses on 20 recent strategy or green planning initiatives undertaken in ten industrial member countries of the Organization for Economic Cooperation and Development (OECD) and in two countries in eastern Europe, and examines the European Union's attempt to develop a sustainable development strategy through its Fifth Environmental Action Programme.

Structured interviews were undertaken with the key individuals involved in and responsible for managing the selected strategies. The aim was to examine, from the particular perspective of these key players, such issues as the objectives that drove each strategy; the processes adopted and why; the problems encountered and how they were overcome; and factors seen as determining success or failure in strategy development and/or implementation.

Another approach would have been to consult widely with all actors and interest groups in each country, i e to conduct a post-strategy audit. The costs of undertaking such a study on an international basis would be large, but such reviews could be carried out in individual countries by local organizations.

The countries and strategy processes covered in this study are listed in Table 1.1. In some countries, more than one national strategy was included.

Table 1.1 Study Interviews: Countries and Strategies

Country	Strategy	Proponent
Australia	National Strategy for Ecologically Sustainable Development (1992)	Environment Strategies Directorate, Dept Environment, Sport and the Territories
Canada	Projet de société (development began '92, final draft June '95)	National Round Table on the Environment and the Economy
Canada	Green Plan (operational 1990–6)	Environment Canada
Denmark	Nature and Environment Policy Plan (initiated June '94, published August '95)	Ministry of the Environment and Energy
European Union	Fifth Environmental Action Programme (initiated March '92, approved Feb '93)	Environment Directorate General, European Commission
France	Plan national pour l'Environnement (1990–)	Ministry of the Environment
France	French Commission for Sustainable Development: key issue studies	French Commission for Sustainable Development (established 1994)
Latvia	National Environmental Policy Plan for Latvia (approved 1995) National Environmental Action Plan (work began November '94, due October '95)	Ministry of Environmental Protection and Regional Development
Netherlands	National Environmental Policy Plan '89; NEPP+ '90; NEPP2 '93	Ministry of Housing, Spatial Planning and the Environment
Netherlands	Action Plan Sustainable Netherlands (1993)	Milieudefensie (Friends of the Earth Netherlands)
New Zealand	*Resource Management Act* (1991); *Environment 2010 Strategy* (adopted July '95)	Ministry for the Environment
Norway	Reports to Parliament on WCED follow-up ('89) and on UNCED ('92)	Ministry of the Environment
Poland	National Environmental Policy (approved by Council of Ministers 1990, accepted by Parliament '91); Implementation Plan to Year 2000	Ministry of Environmental Protection, Natural Resources and Forestry
Sweden	Towards Sustainable Development in Sweden. Government bill (adopted '94)	Ministry of the Environment
Sweden	An Environmentally Adapted Society: Action Programme: Enviro '93 (1993)	Swedish Environmental Protection Agency
UK	Strategy for Sustainable Development (1994)	Department of the Environment
USA	President's Council on Sustainable Development: national sustainable development action strategy (presented to the President March '96)	President's Council on Sustainable Development
USA	National Environmental Goals Project (established 1992; completion due '95)	Environmental Protection Agency

Based on the lessons and principles in Box 1.2, and the dilemmas listed above, a questionnaire was drawn up and used as a basis for conducting each of the structured interviews (see Appendix 1). The questions were grouped under the following main headings:

- How did the strategy get going and why?
- What were/are the main aims and the focus?
- How was/is the strategy being organized and managed?
- Who participated in the strategy?
- What were/are the key factors, issues and problems?
- How were/are problems and conflicts solved?
- How did/does the strategy relate with other strategies?
- What was/is the driving perspective and were/are wider issues dealt with (for example, ecological footprints, transboundary issues)?
- Has the strategy led to a parliamentary and wider debate?

The interviews were conducted between December 1994 and February 1996. All were updated by correspondence with those interviewed prior to publication. Chapters 9 to 21 in Part II of this book are country case studies based on these interviews.

Chapter 2 APPROACHES and STIMULI

Analysis reveals a range of different approaches which have characterized the development and implementation of green plans and strategies in industrial countries. The green plans and strategies covered by this study are compared against these characteristics in Table 2.1.

The Stimulus of UNCED and Environmental Concern

Since the 1972 UN Conference on the Human Environment in Stockholm, there has been increasing acceptance of the importance of environmental issues. The called landmark report of the World Commission on Environment and Development, commonly the Brundtland Commission (WCED, 1987), had a profound influence on governments in the North and South alike. It promoted closer links between environment and development and emphasized issues of social and economic sustainability. The 1992 UN Conference on Environment and Development (UNCED) was organized mainly to respond to the challenges set out in the Brundtland Commission report.

As already noted in Chapter 1, Agenda 21, one of the main accords reached at UNCED, calls on all nations to develop national sustainable development strategies to meet its objectives. Many have done just that, initiating new processes and/or developing new strategies to give national expression to the challenges outlined in Agenda 21.

In some cases, governments responded by developing new strategies, or reports or bills, for Parliaments. For example, immediately after UNCED, Norway prepared a report to Parliament on UNCED and its implications for the country, *Report No 13 to the Storting*, (NorMOE, 1989), while Sweden developed and submitted a bill to Parliament (Towards Sustainable Development in Sweden, SwedMoE, 1994) on implementing the resolutions of UNCED. Intention to publish a *UK Strategy for Sustainable Development* was announced by Prime Minister Major at UNCED and the document was published in January 1994. More recently published strategies include Denmark's *Nature and Environment Policy* (DanMoE, 1995) and the USA's *Sustainable America: A New Consensus for Prosperity, Opportunity, and a Healthy Environment for the Future* (PCSD, 1996).

Table 2.1 Strategy Characteristics Compared

	Australia	Canada Projet	Canada G Plan	Denmark	EU	France PNE	France FCSD
RESPONSE							
Preceded UNCED	yes	no	yes	no	yes	yes	no
Response to UNCED	no	yes	no	yes	no	no	yes
Response to public concern for the environment	yes	no	yes	no	partly	yes	no
GOVERNMENT/PARLIAMENT							
Official government mandate	yes	yes	yes	yes	yes	yes	yes
High-level government commitment	yes	?	yes	yes	yes	yes	yes
Formally reported to Cabinet	yes	no	yes	yes	N/A	?	no
Formally presented to parliament	no	no	yes	?	yes	yes	no
Led to legislative changes	yes	no	yes	?	yes	yes	no
Led to institutional changes	yes	no	?	?	yes	yes	no
FOCUS							
Environment focus	no	no	yes	yes	yes	yes	no
Sustainable development focus	yes	yes	no	no	yes	no	yes
Concerned mainly with federal areas of responsibility	yes	no	yes	no	N/A	no	no
Set target time horizon(s)	no	no	yes	no	yes	yes	yes
Focus on special issues studies	no	no	no	no	no	no	yes
LINKS							
Directly linked to budget process	no	no	yes	no	no	?	no
Linked to mainstream national planning	no	no	yes	yes	no	yes	no
Formal link with other national strategies	yes	no	no	yes	no	no	no
Influenced by regional strategies	no	yes?	yes	yes	no	?	no
Formal link with provincial/state strategies	yes	no	no	no	no	no	no
Promoted/supported local Agenda 21s	yes	?	no	yes	yes	yes	no
Developed independent from Convention strategies	yes	yes	yes	yes	yes	yes	yes
PROCESS							
Independent process	no	yes	no	no	no	no	no
Continuing (cyclical) process	yes	yes?	no	yes	yes	no	?
Cross-govt./inter-departmental process	yes	no	yes	yes	N/A	yes	no
Undertaken exclusively within single ministry	no	no	no	no	mainly	no	no
PARTICIPATION							
Extensive stakeholder participation	yes	yes	no	no	no	no	no
Consultation with industry/NGOs/public	yes	yes	yes	no	some	yes	yes
Mainly internal government exercise	no	no	yes	yes	yes	yes	no

Table 2.1 continued

	Latvia NEPP/NEAP	Netherlands NEPP	NEPP2	Netherlands Action Plan	New Zealand Res Man Act	Envir.2010	Norway Rep 46	Rep 13
RESPONSE								
Preceded UNCED	no	yes	no	yes	yes	no	yes	no
Response to UNCED	?	no	part	part	no	no	no	yes
Response to public concern for the environment	no	yes	no	no	partly	no	no	no
GOVERNMENT/PARLIAMENT								
Official government mandate	yes	yes	yes	no	yes	yes	yes	yes
High-level government commitment	?	yes	yes	no	yes	yes	yes	yes
Formally reported to Cabinet	yes	yes	yes	no	yes	?	yes	yes
Formally presented to parliament	no	?	yes	no	yes	?	yes	yes
Led to legislative changes	no	yes	?	no	yes	?	yes	no
Led to institutional changes	no	?	?	no	yes	?	?	no
FOCUS								
Environment focus	yes	yes	yes	no	yes	mainly	yes	no
Sustainable development focus	no	part	part	yes	partly	partly	part	yes
Concerned mainly with federal areas of responsibility	no	–	–	–	no	no	no	no
Set target time horizon(s)	yes	yes	yes	yes	no	yes	no	no
Focus on special issues studies	no	no	no	no	no	no	no	no
LINKS								
Directly linked to budget process	no	?	yes	no	no	yes	yes	no
Linked to mainstream national planning	no	?	yes	no	yes	yes	yes	no
Formal link with other national strategies	no	no	no	no	no	yes	no	no
Influenced by regional strategies	yes	–	–	no	no	no	yes	no
Formal link with provincial/state strategies	no	–	–	no	no	no	no	no
Promoted/supported local Agenda 21s	no	no	yes	no	no	yes	yes	no
Developed independent from Convention strategies	no	yes	yes	yes	yes	yes	yes	yes
PROCESS								
Independent process	no	no	no	yes	no	no	no	no
Continuing (cyclical) process	yes	yes	yes	no	no	yes	no	no
Cross-govt/inter-departmental process	yes	yes	yes	–	yes	yes	yes	yes
Undertaken exclusively within single ministry	no	no	no	–	no	no	no	no
PARTICIPATION								
Extensive stakeholder participation	no	no	part	no	yes	no	no	no
Consultation with industry/NGOs/public	yes	no	yes	no	–	yes	yes	no
Mainly internal government exercise	yes	yes	no	no	no	no	yes	yes

Table 2.1 continued

	Poland NEP	Sweden gov bill	Sweden Enviro 93	UK	USA PCSD	USA env goals
RESPONSE						
Preceded UNCED	yes	no	yes	no	no	no
Response to UNCED	no	yes	no	yes	yes	no
Response to public concern for the environment	yes	partly	no	no	?	no
GOVERNMENT/PARLIAMENT						
Official government mandate	yes	yes	departmental	yes	yes	set by EPA
High-level government commitment	?	yes	no	yes	?	?
Formally reported to Cabinet	yes	yes	no	yes	not yet	not yet
Formally presented to parliament	yes	yes	no	yes	not yet	? summer 95
Led to legislative changes	yes	yes	no	no	not yet	not yet
Led to institutional changes	?	not yet	no	limited	not yet	not yet
FOCUS						
Environment focus	mainly	yes	yes	no	no	yes
Sustainable development focus	some	yes	–	yes	yes	no
Concerned mainly with federal areas of responsibility	no	no	no	no	?	? yes
Set target time horizon(s)	yes	yes	yes	yes	?	yes
Focus on special issues studies	no	no	yes	no	yes	yes
LINKS						
Directly linked to budget process	no	no	no	no	no	EPA budget
Linked to mainstream national planning	no	yes	no	no	no	no
Formal link with other national strategies	no	no	no	no	no	no
Influenced by regional strategies	yes	some	yes	no	no	? no
Formal link with provincial/state strategies	no	no	no	no	no	no
Promoted/supported local Agenda 21s	no	yes	yes	finance support	no	no
Developed independent from Convention strategies	yes	yes	yes	yes	yes	yes
PROCESS						
Independent process	no	no	no	no	no	no
Continuing (cyclical) process	no	yes	no	no	?	?
Cross-govt./inter-departmental process	some	yes	yes	yes	no	some
Undertaken exclusively within single ministry	mainly	no	no	no	no	mainly EPA
PARTICIPATION						
Extensive stakeholder participation	no	no	no	no	partly	no
Consultation with industry/NGOs/public	some	yes	no	yes	yes	limited
Mainly internal government exercise	yes	yes	yes	yes	no	yes

In other cases, independent initiatives have been established (for example, Canada's *Projet de société* facilitated by the National Round Table on the Environment and the Economy; and the *Action Plan: Sustainable Netherlands* (Milieudefensie, 1992).

Many countries have established their own national Commission for Sustainable Development (CSD), mirroring that of the UN, to address their respective sustainable development issues. According to Silveira (1995), 'Since 1992, over 130 countries have established some form of structure to follow-up Rio agreements and more than 50 countries have initiated official participatory mechanisms to formalize social participation and promote a multi-sectoral dialogue.' Regardless of the names they have been given (council, commission, forum, round table) these national councils for sustainable development generally comprise a 'panel' of appointed persons eminent in their fields (i e academics, experts, business leaders and heads of influential organizations). In the USA, the President's Council on Sustainable Development (PCSD) was established by President Clinton in 1993 and published its national action strategy in March 1996. The French Commission for Sustainable Development (FCSD) was decreed in 1992, but only became active in 1994. Its early work concentrated on preparing reports reflecting on selected key issues for sustainable development. In October 1995, it was announced that a national strategy for sustainable development would be developed by the Ministry of the Environment during 1996 under the guidance of the Prime Minister. The FCSD will also play a full part in this process.

To date, four regional meetings of national councils or similar entities have been held: an Inter-American, October 1994; a European, January 1995; an African, May 1995; and an Asian, June 1995. The European Conference of the National Commissions on Sustainable Development was hosted by France at Courchevel, and attended by representatives from 25 industrial countries. Participants shared experiences and views. The report of the conference (Commissariat général du Plan, 1995) provides a useful synthesis of challenges and issues perceived by CSDs and contains an annex outlining the work of the commissions attending. A useful *Directory of National Councils for Sustainable Development – or Similar Initiatives* is also now available (Earth Council, 1995). It aims to facilitate communication and exchange between such bodies.

While UNCED gave new impetus to addressing sustainable development, many industrial countries were already engaged in some form of green planning to deal with pressing environmental problems. Such initiatives were frequently, in some measure, a response to growing public pressure for action to address serious environmental concerns. For example, in the Netherlands, the first *National Environmental Policy Plan* (NEPP) (VROM, 1989) was initiated in 1987, partly in response to public demand for a more active government role following the Chernobyl accident and because of domestic scandals concerning soil pollution. It was an integrated initiative, shifting from a sectoral to a theme-based approach to environmental planning and management, and dealing with a range of source and process themes. The New Zealand *Resource Management Act* (1991) was developed to rationalize severe inequities in the way environmental management operated across different sectors, partly as a response to concerns raised by environmental and industrial groups. Similarly, the Australian government's initiation of the process, which led to the publication in 1992 of the *National Strategy for Ecologically Sustainable Development* (CoA, 1992a), was to a significant extent triggered by domestic public pressure to sort out issues concerning resource use and environment–development conflicts encountered by successive governments (for example, the construction of the Franklin dam in Tasmania and mining in Kakadu National Park). In Canada, political demand for the *Green Plan* (Government of Canada, 1990a) grew out of increased environmental awareness in the late 1980s, public concern following accidents involving oil spills and PCB fires, and a series of high-profile controversies over the environmental assessment of major projects (Toner, 1994). The ruling Conservative Party sought to turn this environmental concern to its advantage by taking action and gaining electoral support. This issue is returned to in Chapter 7's section on political influences. The European Union's Fifth Environmental Action Programme, also called the 'Towards Sustainability' strategy, was conceived as a response to the report of the Brundtland Commission (WCED, 1987) and in preparation for UNCED.

Many recent green planning initiatives in central and eastern Europe (CEE) have been guided by the Environmental Action Programme (EAP) for Central and Eastern Europe, endorsed in April 1993 by European environment ministers meeting in Lucerne. The EAP aims to assist CEE countries transform their systems of environmental protection, which were strongly influenced by their former centrally-planned economic systems, through the adoption of market instruments to control pollution. In a recent report, the Regional Environmental Centre for Central and Eastern Europe has assessed progress made by 12 CEE countries in adopting national environmental programmes (NEAPs) (REC, 1995). Some key characteristics of these NEAPs are shown in Table 2.2. To a large extent, these NEAPs have been developed to demonstrate commitment to reversing environmental degradation in order to attract investment and donor assistance. The REC report identifies two main and contrasting trends which have emerged that illustrate the position of environmental issues relative to economic and social problems in the CEE countries:

- In some countries (Bulgaria, Croatia, Czech Republic, Hungary, Latvia, Lithuania, Poland and Slovakia), the public and government concern for environmental issues was very high, but has decreased over the past four years and been replaced by concern about economic and social issues. This process is seen as an adjustment in attitude as a result of economic hardship. However, it is not an indication that these countries have solved earlier identified environmental problems.
- In other countries (Albania and the Federal Yugoslav Republic of Macedonia), environmental problems have only recently gained some attention. This is due to the increased availability of information demonstrating the relationship between the environment, human health and the economy. Important environmental issues are still under-evaluated and need to be recognized and addressed properly.

As the report makes clear, 'Both trends point to the complex task that lies ahead as the CEE countries identify and prioritize development goals and environmental issues during the transition period. Their ability to effectively address these environmental issues depends not only on the level of public support, but also on the ability of each environmental administration to act' (REC, 1995).

The development of Poland's 1990 National Environmental Policy (PolMEP, 1990a) preceded the EAP and was the result of a series of events connected with the political changes in the country, growing awareness of the severe environmental damage suffered during the communist period and pressure for radical policy reform.

Table 2.2 Key Characteristics of NEAPs in Central and Eastern Europe

	Albania	Bulgaria	Croatia	Czech Republic	Hungary	Latvia	Lithuania	FYR Macedonia	Poland	Romania	Slovak Republic	Slovenia
1. **Environment as a national priority**												
1990	No	Yes	No	Yes	Yes	Yes	Yes	No	Yes	Yes	Yes	No
1994	Yes	No	No	No	No	No	No	No	No	Yes	No	No
2. **Environmental policy making**												
Environmental frame act	'93	'91	'94	'92	'95	'90	'92	tbe	'80	'89	'93	'93
Environmental strategy document	'93	'92	tbc	'90	'94	'95	tbc	tbc	'91	'92	'93	tbc
3. **Capacity for environmental administration**												
a) individual skills												
technical expertise	No	Yes	•	•	Yes	•	•	Yes	Yes	Yes	Yes	Yes
managerial skills	No	No	•	•	No	•	•	No	No	No	No	No
b) institutional framework												
adjusted functions and structures	Yes	Yes	Yes	Yes	Yes	Yes	Yes	Yes	Yes	Yes	Yes	Yes
special training programmes	No	Yes	No	No	Yes	No	No	No	No	No	No	No
4. **Priority fields for environmental investments**												
1	dws	apc	iwm	apc	iwm	wwt	wwt	dws	apc	dws	dws	apc
2	dwm	dws	dws	tt	iwm	apc	wwt	dws	fm	apc	wwt	dwm
3	wwt	dwm	dwm	wwt	dwm	dwm	tt	idwm	wwt	nhm	wwt	tt
5. **Priority in development of the NEAP components**												
Environmental policy development	3	3	2	3	3	3	1	1	2	1	2	2
Institutional strengthening	1	1	1	1	1	1	1	2	1	2	3	1
Investments	2	2	3	2	2	2	1	2	3	3	1	3
6. **Number of local language EAP copies delivered**	350	700	800	1000	1500	350	378	520	3500	2500	500	1000

Key: apc (air pollution control), dwm (domestic waste management), dws (drinking water supply), fm (forest management), idwm (industrial and domestic waste management), iwm (industrial waste management), nhm (nuclear and other highly hazardous material), tbc (to be completed), tbe (to be enacted), tt (transport, traffic), wwt (waste water treatment)

Note: 1. New will be enacted in 1995

Source: REC (1995)

Other pre-UNCED strategies had less to do with public environmental concerns. For example, the French *Plan national pour l'Environnement* (FrenchMoE, 1990) was developed mainly in order to strengthen (through reorganization) the structure of the Ministry of Environment and to give weight to environmental policy. Norway's preparation of Report No 46 to the Storting, 1988, as a follow-up to the World Commission on Environment and Development, probably had much to do with Mrs Brundtland (the Norwegian Prime Minister) having chaired the commission. Work on *Enviro '93* (SwedEPA, 1993) the 1993 action programme of the Swedish Environmental Protection Agency, began in 1991 as an internal agency assessment of environmental trends, which was undertaken to show how individual sectors would need to respond in developing their own sectoral environmental action plans and programmes.

However, many of the strategies and equivalent green planning exercises that commenced before UNCED were undoubtedly subsequently influenced by it in one way or another. Agenda 21 has also influenced revisions of past strategies or strategies that effectively build on past initiatives. For example, in New Zealand, development of the *Environment 2010 Strategy* (NZMfE, 1994a) was initiated in order to set out a broader vision following the 'micro' reforms introduced by the 1991 *Resource Management Act*. In preparing for their inputs to this strategy and for related initiatives, most government departments analysed the significance of Agenda 21 for their policies and activities, even though drawing only broad implications. Equally, development of the Netherlands' second *National Environmental Policy Plan* (NEPP2) (VROM, 1993a) was started in mid-1992 and was able to take into account the messages and lessons in Agenda 21.

Chapter 3 FOCUS of GREEN PLANS and STRATEGIES

It is increasingly accepted that sustainable development is about achieving a quality of life that can be maintained for many generations because it is:

- *socially desirable*, fulfilling people's cultural, material and spiritual needs in equitable ways;
- *economically viable*, paying for itself, with costs not exceeding income; and
- *ecologically sustainable*, maintaining the long-term viability of supporting eco-systems.

In a recent paper discussing participation in strategies, Bass et al (1995) comment that:

> *Sustainable development is a challenging social process. Decisions need to be made about the relative rights and needs of present and future generations. Choices have to be made between priorities at local, national and, indeed, global levels. The different objectives of society — social, economic and environmental — need to be integrated where possible, and traded-off where they are incompatible. Institutional and individual roles and responsibilities have to change, so that new patterns of behaviour will foster sustainable development.*

However, among the strategies and green planning initiatives covered in this study, even among those purporting to focus on sustainable development, it is not always easy to determine whether any genuine attention has been paid to the social dimensions, or whether — and how effectively — the economic issues relating to environmental management and social matters have been addressed. Clearly, all have dealt in considerable depth with natural resources, conservation and environmental issues and problems in their countries.

Of the 21 case studies covered in this report, only eight are concerned specifically with defining national agendas for and responses to the challenge of sustainable development *per se*, while a further five focus partly on sustainable development issues. Over half are concerned exclusively or dominantly with environmental issues and often they have been strongly influenced by a need to respond to growing public pressure to 'do something' about the environment. While these distinctions may be useful, in reality there is a continuum of approaches, and environmental plans and strategies can, in time, evolve to become sustainable development strategies.

Sustainable Development Strategies

As might be expected, those strategies or equivalent initiatives concerned predominantly with sustainable development were almost all undertaken after UNCED (1992) and were mainly stimulated by commitments under Agenda 21 — either as a formal government response in the form of a report or bill (for example, Sweden, Norway and the UK); through the formal establishment of a national Commission for Sustainable Development (for example, France and the USA); or as an independent response to UNCED (for example, the Canadian *Projet de société*). By comparison, the independent *Action Plan: Sustainable Netherlands*, was developed by Milieudefensie as part of its preparations for UNCED.

An exception was the Australian *National Strategy for Ecologically Sustainable Development* (NSESD), initiated in 1990. The World Conservation Strategy (IUCN/UNEP/WWF, 1980), Australia's 1983 conservation strategy and the Brundtland Commission report (WCED, 1987) progressively stimulated thinking about sustainable development in Australia. The strategy can also be traced to a cooperative approach between the Farmers' Federation and the Australian Conservation Foundation to tackle land degradation — Landcare — and an outline proposal to the federal government by these two organizations for a strategy approach.

Australia's NSESD and Canada's *Projet de société* are good examples of strategies adopting a holistic approach. The NSESD, while concerned centrally with 'maintaining the ecological processes on which life depends', tries equally to deal with 'enhancing individual and community well-being and welfare by following a path of economic development that safeguards the welfare of future generations', and aims to 'provide for equity within and between generations' (CoA, 1992a) (see also Boxes 9.2 and 9.3). This broad focus resulted in some 500 issues, grouped into sectoral and crosscutting areas, being covered. But there was no actual focus on options that could make a real difference such as influencing the greening of the economy — micro-economic reform is still treated separately; and environmental 'bottom lines' are not treated systematically, as in New Zealand's *Resource Management Act* (1991). Ironically, by embracing 'all the issues' in the NSESD (as advocated by a sustainable development approach), sustainability has been found difficult to 'policy manage' and conventional institutional arrangements are severely taxed. As already noted, there has also been criticism of Australia's failure to legislate to integrate the fundamental issues of sustainable development into federal law (Scanlon, 1995).

As indicated above, a key task in addressing sustainable development is making trade-offs where objectives have been found to be incompatible. Few strategies have effectively dealt with this issue, and this is perhaps not surprising since there are still few tried and tested tools to help in this process. However, in the draft strategy of the Canadian *Projet de société*, innovative 'choicework' tables around basic human needs (for example, air, water, food, mobility) are provided to assist stakeholders to reach innovative solutions (see Boxes 10.2 and 10.3). Choicework is defined as 'sorting out choices, weighing pros and cons, and beginning making the difficult trade-offs' (Projet de société, 1995b). The tables represent an attempt to compare expert and public perceptions of various issues with a view to finding a method of bridging the gap between experts and the general public on a range of sustainability issues. They also identify areas of conflict and levels of consensus to show where immediate progress can be made and where more consensus-building is needed.

The new national commissions for sustainable development have adopted different approaches. The French CSD focused initially on considering various priority themes and issues (for example, in 1995 the focus was on indicators for sustainable cities and urban ecological issues). By comparison, in the USA, the President's Council on Sustainable Development has issued its report to the president entitled *Sustainable America: A New Consensus for the Future* (PCSD, 1996). After two and a half years of inquiry, observation and discussion, 25 leaders from business, government, environmental, civil rights and native American organizations reached unanimous agreement. Their report includes a vision statement and fundamental beliefs on sustainability; indicates changes in decision-making at all levels — government, business and community, and both institutional and individual — that must occur to achieve sustainable development; and provides scores of wide-ranging recommendations and actions with which to implement them. At the heart of the council's recommendations is the conviction that economic, environmental and social equity issues are inextricably linked and must be considered together. It is stressed that, to achieve sustainability, institutions and individuals must adopt this new way of thinking; and that some things — jobs, productivity, wages, capital and savings, profits, information, knowledge, and education — must grow, and others — pollution, waste and poverty — must not.

In Norway, after UNCED, a CSD was established and chaired by the Prime Minister as a forum in which to exchange ideas on important environmental issues before policy is finalized. Seven meetings were held and issues such as climate change and the employment effects of environmental measures were discussed. But the commission has been inactive for over a year now. In Sweden, the national CSD was established in 1994 with the aim of developing a basis for a Swedish report to the forthcoming 1997 special session of the UN General Assembly on progress since UNCED. The Swedish CSD will also coordinate follow-up to Government Bill 1993/4:111 (April 1994) on implementing the resolutions of UNCED.

The Swedish bill is based on the overall agenda of sustainable development and is viewed as a 'quantum leap' from the early 1990s when 'green attention' was given only to the environment. It lays down general guidelines for development in Sweden within the various problem areas and sectors of society. Together, they form a national strategy for sustainable development and thus constitute a Swedish Agenda 21 based on environmental goals and priorities already adopted by the Riksdag (Parliament), *inter alia* in terms of the development of an 'ecocycle' society, climate questions, biological diversity, forestry policies, matters currently at the preparatory stage, such as an environment code, and questions concerning chemicals and the environmental debt. A report on the state of the environment — the government's annual environmental report to the Riksdag — is appended to this bill, summarized in Box 18.1.

The independent *Action Plan: Sustainable Netherlands*, 1993, developed by Milieudefensie, uses an innovative concept of 'environmental space' (see Box 3.1) to examine

Box 3.1 ENVIRONMENTAL SPACE AND ECOLOGICAL FOOTPRINTS

In its 'Action Plan for a Sustainable Netherlands' published in April 1992, the Dutch Friends of the Earth, Milieudefensie, made a rough calculation of available per capita global carrying capacity (or 'environmental space') for key energy, water, raw materials and arable land resources. It then identified the cuts in current consumption levels necessary in the Netherlands to return to sustainable levels by 2010: these ranged from 40 per cent for fresh water to 80 per cent for aluminium use. As a result of these and other calculations, the Dutch government was one of the few at Rio to acknowledge that it could only sustain its lifestyle by exploiting the carrying capacity of other countries (VROM, 1991).

The parallel concept of 'ecological footprints' was coined by William Rees to describe the tendency of urban areas 'through trade and natural flows to appropriate the carrying capacity of distant elsewheres'. Looking specifically at the Lower Fraser Valley of British Columbia, Canada, Rees found that the land area required to support the community (in other words, its 'ecological footprint') was at least 20 times the land it occupies. Looking at the issue from a Southern perspective, Anil Agarwal at the Centre for Science and Environment in India has estimated that the total biomass currently exported from the developing world to industrialized countries is ten times greater than during the colonial period (Weizsäcker, 1994). These exports of carrying capacity do not necessarily pose a problem if they are drawing on true ecological surpluses, and if enough remains for meeting local needs. Currently, there is no guarantee that trade flows are really based on these principles.

what a sustainable Netherlands would look like in 2010 and what steps would be needed to achieve it. It is not really an action plan, but sets goals for action, although not defining how to implement them. The parallel concept of 'ecological footprints' (see Box 3.1) is raised cautiously in the *UK Strategy for Sustainable Development* (HMSO, 1994a) as a 'difficult issue to think about', but without resolving the government's view.

The aim of Norway's Report No 13 to the Storting on UNCED 1992 (NorMoE, 1992b) was specifically to review policies in the light of the relatively demanding texts agreed at the conference, particularly Agenda 21.

The UK's *Strategy for Sustainable Development* (HMSO, 1994a) sets out the government's perspective on this issue. According to Stevens (1995), it 'sets out to examine not only environmental protection, but the essential relationship between this and economic activity'. The government took the view that social issues need not be a prominent theme, and that they would be dealt with as they 'ran in and out' of the issues of primary concern linked to the above focus. Whether or not discussions during the development of the strategy dealt effectively with the social dimensions is not clear. But the strategy document itself is relatively weak in its consideration of the social interface between environmental and economic issues. It provides principles for sustainable development, identifying the main 'problems and opportunities' in each sector over the next 20 years. The process also reviewed and led to the development of institutional arrangements for pursuing sustainable development policies (i e the establishment of an independent panel to advise government, a national round table representing all sectors, and a 'Going for Green' campaign to promote sustainable development at the individual level).

Environmental Strategies and Plans

The Canadian *Green Plan* (1990–6) (Government of Canada, 1990a) — arguably the 'mother' of green planning — preceded UNCED and was never intended to be a vehicle for sustainable development. Its aim was to tackle pressing environmental problems and concerns, although Environment Canada staff have commented that it was undertaken 'within a context of sustainable development'. The focus is very much on decision-making in relation to environmental problems and on an action plan to address specific issues (for example, climate and fisheries). However, a follow-up process is now under way and here the aim is to 'integrate environmental and economic considerations, and the social aspects of sustainable development into government decision-making as our understanding of these aspects grows' (see Chapter 10). New legislation requires all federal ministers to table, in Parliament, departmental sustainable development strategies by the end of 1997. These strategies are to be updated and tabled in Parliament every three years.

In the Netherlands, the three *National Environmental Policy Plans* published to date are: NEPP (VROM, 1989), NEPP+ (VROM, 1990) and NEPP2 (VROM, 1993a). These are part of a periodically revised environmental strategy which, like the Canadian *Green Plan*, are set in a sustainable development context. But the main aim of the first NEPP was to deal with the national pollution problem. It marked a shift from the compartmentalized, effect-oriented approach of the 1970s and 1980s (focusing on soil, water and the like) to an integrated policy approach through eight environmental source and process themes:

- climate change (global warming and depletion of the ozone layer);
- acidification (acid deposition on soil, surface water and buildings);
- eutrophication (excessive nutrient build-up in surface water);
- diffusion (uncontrolled spread of chemicals — dispersion of toxic waste and hazardous substances — through environmental media);
- waste disposal (waste processing, waste prevention, reuse and recycling);
- disturbance (nuisance caused by noise and odour);
- groundwater depletion (leading to habitat change or insecurity of water supply); and
- squandering (inefficient use of natural resources, energy and raw materials).

The 1989 NEPP set out the Dutch government's environmental agenda and created the momentum for many other groups in society to develop their own plans and programmes (VROM, 1989). The function of the 1993 NEPP2 was to follow through these many initiatives and to ensure that their objectives are realized and NEPP targets are met (VROM, 1994a). However, there was less public concern about the environment during the preparation of NEPP2, when there was more concern about issues such as the European Union, immigration, and the level of criminality.

After UNCED, the Danish Ministry of the Environment (MoE) instituted a process of broad rolling strategic planning. The two key elements of environmental planning in

Denmark will be an environmental progress report (state of the environment), to be drawn up at appropriate intervals; and a White Paper on the environment, describing overall priorities, targets and specific initiatives in respect of future environmental action, and drawn up with a maximum frequency of four years. The first of the White Papers is the new Danish *Nature and Environment Policy Plan* (DanMoE, 1995). MoE staff indicated that this plan is intended to represent Denmark's national sustainable development strategy, but it is essentially environmental in scope and prepared as a background for holistic planning. It does not deal with social and economic strategies.

France's *Plan national pour l'Environnement* (PNE) (FrenchMoE, 1990a) is similarly environmental in scope, dealing with issues such as waste and pollution, and developing 'horizontal' approaches such as eco-taxes. A major impetus was to provide a vehicle for reform in public administration of environmental management (strengthening the structure of the Ministry of Environment) and to focus on the environmental responsibilities of other ministries.

In bringing together the problems and policies concerning the environment in a single policy document, Norway's *Report No 46 to the Storting* (NorMoE, 1989), a follow-up to the Brundtland Commission report, also set out the responsibilities of individual sectors for achieving environmental aims. In addition, Report 46 introduced the idea of a separate 'green budget' (see Box 6.4). The report is an example of a green planning exercise in which considerable emphasis is placed on cost-effective and economic measures, as appropriate, to reach the aims. These considerations have been discussed by Paul Hofseth (1993) (see Box 16.2).

Following the break-up of the former Soviet Union, a major issue in many eastern European countries is environmental 'clean up' and protection, and strategies tend to be focused around this priority agenda. In Poland, the *National Environmental Policy* (NEP) (PolMEP, 1990) recognizes the need to focus on 'green reconstruction' of particular sectors of the economy that pose serious threats to the environment, for example energy, industry, transportation, mining, agriculture and forestry, and identifies the need for a system of authority and responsibility in central and local government. The NEP departs from the former narrow and centrally-controlled approach towards environmental protection and incorporates broader goals of achieving sustainable development.

In Latvia, the *National Environmental Policy Plan* (NEPPL) and the NEAP were prepared not only to provide a base for sustainable development, by integrating the environment into all key sectors, but also to raise environmental awareness within key sectoral ministries. The NEPPL outlines four environmental policy goals, ten priority problems and ten environmental policy principles.

The strategy initiatives of the environmental protection agencies (EPAs) in Sweden and the USA are both, predictably, focused on environmental concerns. The former, *Enviro '93* (SwedEPA, 1993) seeks to shift environmental management in Sweden from a media-based approach (for example, focusing on air, land and sea) to a sector-oriented one with each sector assuming its own responsibility and developing its environmental action plan and programme. This is a common trend in almost all recent green planning. In the USA, the EPA's Environmental Goals Project is setting goal statements around 14 sets of environmental targets (for example, for clean outdoor air, safe drinking water, public awareness and participation).

Special Focus Studies

All strategy processes have usually involved consideration of a wide range of issues, and often the preparation of special subject or theme papers to facilitate debate. In some cases, however, the process has been dominated by a focus on special studies or tasks. Particularly notable in this regard is the work of the French CSD and of the US President's Council on Sustainable Development, the preparations for Sweden's *Enviro '93* action programme and the US Environmental Protection Agency's Environmental Goals Project.

The French CSD is able to select its own work areas, make agenda proposals and seek solutions. It initially focused on investigating priority themes rather than preparing a national sustainable development strategy (see discussion on 'focus' in Chapter 12).

In the USA, the President's Council on Sustainable Development organized its work around eight main task forces, focusing on particular themes, for example, eco-efficiency, sustainable agriculture, energy and transportation. The task forces each developed goals and recommendations for the council to consider.

In developing *Enviro '93* (SwedEPA, 1993) Sweden's Environmental Protection Agency established project groups to concentrate on developing background reports on key areas of environmental concern (for example, acidification and metals in the environment) and a subsequent series of sector-based action plans (for energy, traffic, industry and forestry). These reports and plans were later consolidated to produce *Enviro '93*.

The US EPA has concentrated its Environmental Goals Project around 13 key goals (for example, clean air and safe drinking water). Each goal statement will include:

- a problem statement;
- a general goal for the condition of the environment;
- specific targets for 2005;
- actions to achieve them;
- current status and trends;
- government (federal, state, local) responsibilities; and
- implications for society.

The independent *Action Plan: Sustainable Netherlands*, prepared by Milieudefensie (1992) was organized around consideration of the concept of 'environmental space' (see Box 3.1).

Chapter 4 DURATION, TIME FRAMES, MANDATES and MANAGEMENT APPROACHES

Duration of Development Process

The amount of time taken (or allocated) to develop strategies in industrial countries has varied between about six months and three years, varying in some cases according to the process involved (Table 4.1). Those adopting a participatory approach with broad stakeholder involvement have tended to take two to three years (for example, Canada's *Projet de société*, New Zealand's *Resource Management Act*). In contrast, those prepared as a form of government report following established procedures (such as a bill to Parliament), have generally been completed in less than a year (for example, Norway's 1992 Report No 13 to the Storting on the UNCED follow-up, and Sweden's 1993 'Towards Sustainable Development' government bill).

Most strategies have adopted or been based on some time frame for the visions they contain or for implementation of actions. In a number of cases, the strategy goals, objectives and targets are set in relation to several time frames. Table 4.2 lists the time frames of the strategies studied.

Mandates and Terms of Reference

Most of the strategies initiated by governments have had some form of official mandate or terms of reference (ToR). In some cases, this has been a formal instrument, as in the USA where the President's Council on Sustainable Development was established as a result of an executive order by the president. In other cases, ToR have been issued by the head of government. For example, in Australia, the *National Strategy for Ecologically Sustainable Development* (COA, 1992a) was initiated by the establishment of working groups. The ToR for these working groups were set out in 'charter letters' from Prime Minister Bob Hawke to the chairpersons of the three subcommittees into which

Table 4.1 Duration of Strategy Preparation

Australia, NSESD	*c* 18 months (August 1990–Dec 1992)
Canada, Projet de société	*c* 3 years (1992–5)
Canada, Green Plan	18 months (1989–90)
Denmark, Nature and Envir. Policy Plan	1 year (June 1994–June 1995)
European Union, Fifth EAP	18 months (mid-1991–February 1993)
France, PNE	6 months (1989)
France, CSD	open process (1994–)
Latvia, NEPPL	18 months (December 1993–April 1995)
Latvia, NEAP (NEPPL implementation prog)	*c* 1 year (November 1994–October 1995)
Netherlands, NEPPs	NEPP *c* 2 years (1987–9) NEPP+: *c* 5 months (revision in 1990) NEPP2: *c* 1 year (mid-1992–Sept 1993)
Netherlands, Action Plan: Sust Neth	2 years (1991–3)
New Zealand, RMA	3 years (1988–91)
New Zealand, Envir. 2010	1 year (October 1994–September 1995)
Norway, Rep 46	2 years (1987–9)
Norway, Rep 13	6 months (1992)
Poland, NEP	< 1 year (1990)
Sweden, government bill	< 1 year (spring–December 1993)
Sweden, Enviro '93	*c* 18 mths (autumn 1991–summer 1993)
UK, Sustainable Development Strategy	*c* 9 months (spring 1993–January 1994)
USA, PCSD	*c* 2.5 years (June 1993–February 1996)
USA, Envir Goals Project	*c* 3 years (1992–summer 1995)

these working groups were organized. To guide the working groups, the letters set out fundamental goals and principles to which the then government was firmly committed. Similarly, in France, the ToR for the *Plan national pour l'Environnement* (French MoE, 1990a), were effectively set out in a 1989 letter from the Prime Minister to the Minister of the Environment, Brice Lalonde. Initiation of the Dutch *National Environmental Policy Plan* (NEPP) in 1987 was also heavily guided by the Prime Minister because it was a top political priority. ToR for the follow-up NEPP2 were set by the five key ministries involved. At first it was difficult to secure agreement on them and the matter was resolved by the Cabinet after 'detailed discussions'. Other strategy initiatives have also been initiated by decisions of cabinets, for example, Report 46 of 1988 and Report 13 of 1992 to the Norwegian Storting.

However, some strategies initially had no set ToR and civil servants in the lead agencies set the agenda. Environment Canada (the Ministry of Environment) at first had to interpret government thinking in designing its work on the 1990 *Green Plan*. The Minister of the Environment subsequently issued instructions to develop Canada's position on environmental policy. In Latvia, no written ToR within which to develop the *National Environmental Policy Plan* were set, for the process was agreed during lengthy discussions within the Ministry of Environmental Protection and Regional Development, which included the minister. However, following a request from the

***Table 4.2** Strategy Time Frames*

Australia, NSESD (1992)	No set time frame
Canada, Projet de société (1995 draft)	No set time frame
Canada, Green Plan (1990)	6 year implementation prog (1990–6), but many targets had 10-year perspective
Denmark, Nature & Envir. Policy Plan (1995)	Short-term visions Long-term visions
European Union, Fifth EAP (1993)	Long-term perspective, with short-term performance targets for 2000
France, PNE (1990)	10 years (1990–2000)
France, CSD (established 1994)	No set time frame
Latvia, NEPPL (1995)	Long-term (20–30 years)
Latvia, NEAP (NEPPL implementation prog)	Short-term (1–5 years, from 1995)
Netherlands, NEPP (1989)	Meet sustainability goal by 2010 (or 2000 for some objectives)
NEPP+ (1990)	Ditto
NEPP2 (1993)	Ditto, plus state of environment report (1993–2010), 4-year plan, and annual rolling 3-yearly environment programme
Netherlands, Action Plan: Sust Neth (1992)	Vision of sustainable Netherlands in 2010
New Zealand, RMA (1991)	No set time frame
New Zealand, Environment 2010 (1995)	Rolling review every 4–5 years
Norway, Rep 46 (1989)	No set time frame
Norway, Rep 13 (1993)	No set time frame
Poland, NEP (1990)	Short-term priorities (3–4 years) Medium-term (10 years, until 2000) Long-term (20–25 yrs, until 2020) Implementation Plan to Year 2000
Sweden, government bill (1994)	Actions 1994–7, + long-term objectives
Sweden, Enviro '93	Perspective to 2000
UK, Sustainable Development Strategy	20 years (to 2012) (longer for some issues)
USA, PCSD	?
USA, Environmental Goals Project	Targets for 2005

Dutch government, which funded some seminar costs, a project plan (in English) was prepared. In Sweden, civil servants drew up ToR for a government bill (1993/4/111, *Towards Sustainable Development in Sweden*), in cooperation with political advisers in the Ministry of Environment. Similarly, in embarking on the development of New Zealand's *Resource Management Act* 1991, staff of the Ministry for the Environment first developed a proposal, which was then approved by the Cabinet before ToR were set. For the subsequent *Environment 2010 Strategy*, the New Zealand minister for the environment presented a proposal to the Cabinet Committee on Enterprise, Industry and Environment, which endorsed draft ToR. No ToR were set for developing the *UK Strategy for Sustainable Development* (HMSO, 1994a) apart from a broad mandate given by the Prime Minister in his speech at UNCED, and civil servants initiated departmental discussions on how to approach the task. However, ToR were established for various follow-up initiatives.

Green planning initiatives undertaken as projects by specialist government agencies have usually set their own internal ToR. The Swedish Environmental Protection Agency set out ToR for developing its action programme, *Sweden: An Environmentally Adapted Society* (SwedEPA, 1993) in a project plan which was approved by the Ministry of Environment. In the same way, the US Environmental Protection Agency established its own ToR for its National Environmental Goals Project.

Strategies undertaken independently of government have also sometimes published ToR. Those for Canada's *Projet de société* were set out in a 'prospectus', which described the work as a 'multi-stakeholder coalition acting as a catalyst to help promote Canada's transition to a sustainable future' (Projet de société, 1994a).

In most countries, the leading ministry (usually the Ministry of Environment) prepared a discussion paper or description of the initiative; for example, in launching the development of the *Danish Nature and Environment Policy* (DanMoE, 1995) the Ministry of the Environment circulated a 30-page discussion document in the form of a pro-forma ('dummy version') outlining a possible content.

Strategy Management Processes

Few strategy documents describe the process by which they were developed. A clear exception is Australia's *National Strategy for Ecologically Sustainable Development* (CoA, 1992a), which includes a clear outline of the history of the process from initiation to implementation. The process followed for each of the strategies discussed in this study are described in Chapters 9 to 21.

Analysis shows that the development and management of government-driven strategy processes in industrial countries have involved a range of mechanisms. Apart from the issue of participation, which is discussed separately in Chapter 5, these mechanisms include:

- ensuring that primary responsibility to 'manage' the process lies with a 'core team' in an environment ministry or department;
- establishing a senior-level steering committee or group (intergovernmental in some federal countries);
- operating a cross-government process (sometimes acting as a network) with discussions about, negotiations with and contributions from (in the form of sector inputs and draft sections) a wide range of departments or agencies;
- establishing a council or forum of senior figures with representation from industrialists, business leaders, academics, scientists, experts and NGOs.
- establishing working groups or task forces (comprising government officials, experts and academics) to focus on specific issues or themes. They will undertake analyses, identify problems, goals, solutions and targets, and make recommendations (sometimes through special background documents and reports — sometimes by publishing separately from the strategy);
- establishing an advisory group or round table (of distinguished experts) as a forum for discussions and to consider issues;

- holding informal meetings or discussions with representatives from industry, businesses and NGOs;
- releasing draft working papers or draft strategies for public comment;
- publishing public responses and comments;
- holding seminars, workshops, and provincial and public meetings;
- referring difficult issues and conflicts to a high-level committee of senior staff or sometimes ministers;
- putting out a final draft strategy prepared by a 'drafting team' — comprised either of Ministry of Environment or agency staff, or involving several representatives from several agencies. Sometimes the entire strategy is written by an internal departmental/agency team;
- having the draft strategy reviewed by an 'external' group or panel;
- submitting the draft strategy for approval to the Cabinet or to a Cabinet sub-committee, after which it is tabled in Parliament.

Strategies undertaken independently of governments (for example, Canada's *Projet de société*) have involved additional and more innovative approaches, including:

- a secretariat performing a 'facilitating' role rather than coordinating/directing affairs;
- participatory stakeholder round tables and assemblies deciding directions and taking decisions; and
- inputs by volunteers.

Chapter 5 PARTICIPATION in GREEN PLANS and STRATEGIES

Agenda 21 describes a national sustainable development strategy as an approach in which participation is integral, with governments enabling and people managing:

> *National strategies, plans, policies and processes are crucial in achieving this. ... The strategy should build upon and harmonize the various sectoral economic, social and environmental policies and plans that are operating in the country. ... Its goals should be to ensure socially responsible economic development while protecting the resource base and the environment for the benefit of future generations. It should be developed through the widest possible participation.*
>
> (UNCED, 1992)

In a recent study of participation in strategies for sustainable development, Bass, Dalal-Clayton and Pretty (1995) note that successful past strategies appear to have been participatory in nature and, conversely, those that appear to be going nowhere — even though the documentation may look good — frequently have been characterized by a lack of participation. They argue that:

> *Science-based and interdisciplinary approaches are helpful for defining social, environmental and economic trade-offs, but are not sufficient. These kinds of trade-offs are value judgements. They need to be made with the participation of both 'winners' and 'losers', so that some sort of agreement and commitment is reached on the outcome. A people-centred approach is needed as a complement to the science-based approach. Recognizing this, many strategies have built in some elements of participation, in an* ad hoc *manner. Although often without adequate resources and professional skills, these efforts tend to have paid off, and invariably strategies have recommended greater participation in their implementation and further iteration.*

They also propose a typology of participation in national policy-making and planning (Box 5.1).

Box 5.1 A TYPOLOGY OF PARTICIPATION IN POLICY PROCESSES AND PLANNING

1 **Participants listening only** — *for example, receiving information from a government PR campaign or open database.*
2 **Participants listening and giving information** — *for example, through public inquiries, media activities and 'hot-lines'.*
3 **Participants being consulted** — *for example, through working groups and meetings held to discuss policy.*
4 **Participation in analysis and agenda-setting** — *for example, through multi-stakeholder groups, round tables and commissions.*
5 **Participation in reaching consensus on the main strategy elements** — *for example, through national round tables, parliamentary/select committees, and conflict mediation.*
6 **Participants involved in decision-making on the policy, strategy or its components.**

With each level, participation may be:

- narrow *(few actors); or*
- broad *(covering all major groups as well as government)*

(Bass, Dalal-Clayton and Pretty, 1995)

The experience of industrial countries in undertaking green planning and sustainable development strategy processes reveals a wide variety of approaches to the issue of participation. While the majority have exhibited, to a greater or lesser extent, some form of consultation with industry, business interests, academics, special interest groups, NGOs and the public, only a few have involved any serious attempt to involve stakeholders in a genuinely participatory way. Industrial country strategies can, therefore, be crudely categorized as being (a) participatory (categories 4–6 in Box 5.1), (b) strictly consultative (category 3), or (c) dominantly internal to government (categories 1–2).

Participatory Approaches

Some green planning initiatives and sustainable development strategies have adopted a relatively open process and have attempted to involve not only all relevant government departments and agencies and local authorities, but also to engage meaningfully with interest groups, stakeholders, NGOs and the public on aims, content, direction and implementation. In some cases, this has involved a major effort to 'consult' as widely as possible; in others, it has actively involved many actors in the design of the strategy process and even in making major decisions. There is no single model that is common to all. Each has experimented with different approaches and some mechanisms appear to have been more successful than others. Examples include:

- stakeholder round tables;
- providing financial support to enable NGOs to become involved;
- funding NGOs to undertake commissioned work;
- involving target groups; and
- adopting traditional approaches (for example, for meetings).

In Australia, the development of the *National Strategy for Ecologically Sustainable Development* (CoA, 1992a) was initiated in 1990 by the establishment of nine ESD working groups with wide representation, including government officials, industrialists, environmentalists, unionists, welfare officers and representatives of consumer groups, to examine sustainability issues in key sectors of industry. Their purpose was to provide advice on future ESD policy directions and to develop practical proposals for implementing them. Community consultation formed an important part of this process, with a series of one-day consultation forums being held around Australia to discuss mechanisms for integrating economic and environmental concerns, and provide opportunities for broader community comment on the working groups' interim reports.

In May 1992, the heads of state and territory governments released a draft of the strategy to promote discussion and obtain community views on possible future policy directions. This was primarily in recognition of the nature, range and significance of many of the issues covered by the recommendations of the ESD's working group reports. The draft strategy was subsequently released by the Prime Minister on 30 June 1992 and a two-month period was set aside for public comment. Over 200 submissions were received in that period. The majority of these advocated accepting the final recommendations of the ESD working groups and chairs, as well as the mechanisms for implementing them. However, they wanted clearer identification of the priorities and of the agencies responsible for implementation, and elucidation of the linkages between the strategy and other government policies and initiatives. The changes in structure and content in finalizing the strategy were largely in response to these comments. The intergovernmental ESD steering committee also found the submissions a valuable source of information on the broader community's priorities, and utilized them in helping to determine final policy positions.

In developing New Zealand's *Resource Management Act* (RMA), 1991, a massive effort was made to involve the public through public meetings, seminars, free phone-ins and written submissions. All comments received were entered onto a database to prepare a profile of issues. All papers submitted to government on the RMA highlighted where stakeholder views accorded or differed from proposals being made. A special 'stream' for Maori consultation was established. This involved traditional-style meetings (*hui*) with Maori organizations throughout the country to explain the RMA process and to secure views and opinions. Funds were made available to enable NGOs to engage in the process and some NGOs undertook commissioned work. The cost of participating is a common problem faced by NGOs in many countries. Unlike industry and the business sector, NGOs tend to lack the resources to enable them to take part effectively in strategy processes.

In the USA, the President's Council on Sustainable Development (PCSD) operated mainly through a series of task forces, each focusing on a particular issue or area. Each task force adopted its own approach to participation, using different processes and work places to match the nature of its scope of work. For example, the Sustainable

Communities Task Force decided to emphasize a 'grass-roots' approach and involve activists in this movement in all its deliberations. Council meetings or workshops were held in communities that had done pioneering work to realize 'sustainability' in their social, economic and environmental life. Other task forces functioned more in a 'consultation' mode.

The Canadian *Projet de société* — which was started in 1992 as an independent multi-stakeholder initiative to prepare a national sustainable development strategy — arguably represents the most participative national-level strategy process so far attempted. The strategy was developed as a consensus process involving over 80 businesses, and government and independent organizations, in a National Stakeholders' Assembly (Box 5.2). It was designed explicitly to be transparent, inclusive and accountable. Each partner and sector was encouraged to identify and take responsibility for its own contributions. Dialogue and cooperation were considered essential to problem-solving, and shared visions were a keystone. The secretariat for the 'projet' was based in the National Round Table on the Environment and the Economy (NRTEE).

The round-table approach pioneered by Canada has been taken up by many other countries (for example, the UK in implementing its strategy for sustainable development). The recent experiences of Canada in using this approach have been reviewed by the NRTEE (Projet de société, 1995a). Various points of view were put forward on the pros and cons of multi-stakeholder processes and their implications (this emphasis on different perspectives is, itself, part of the round-table approach). The analysis is summarized in Box 5.3.

Consultative Approaches

Most of the green planning and strategy processes in industrial countries have adopted a 'consultative' approach (categories 2 and 3 in the participation typology in Box 5.1). The consultation has, however, often been quite extensive. Though Canada's *Green Plan* (Government of Canada, 1990a) is frequently cited as one of the most far-reaching, government-led consultative activities, its consultation aspect was actually initiated quite late on in the process (see Box 5.4).

According to Environment Canada officials interviewed during this study, there was a deliberate decision that the government would 'own' the Green Plan and its products rather than be a 'partner' along with others in a wider process. However, the government did aim to 'involve' others. This decision was taken at a time when the 'public was calling for "leadership" from the federal government'. Detailed discussions and negotiations were held with all relevant government departments and agencies. Ministers of the main departments were involved as a committee throughout. In addition, a Green Plan coordinating committee (mainly for implementation) was established in which other government departments participated. The consultation process eventually initiated by the Cabinet is discussed in a government booklet, *A Framework for Discussion on the Environment* (Environment Canada, 1990), and in *A Report on the Green Plan Consultations* (Government of Canada, 1990b). The consultations that followed between April and August 1990 were extensive. Toner (1994) records that 'information sessions were held in 30 cities, two-day multi-stakeholder workshops were organized in 17 centres and a final session was convened in Ottawa in August'. Environment Canada estimate that 10,000 people were part of the process in one way or another — attending

Box 5.2 CANADA'S PROJET DE SOCIÉTÉ

The Projet de société recognizes several necessities: that the transition to sustainability is a collective responsibility of all Canadians; that all levels and sectors of society must be engaged in identifying and implementing the necessary changes; and that new institutional models and processes are needed to achieve a common purpose and course of action. These involve partnerships and networks.

Five Canadian organizations came together to organize a first national stakeholder meeting in November 1992: the Canadian Council of Ministers of the Environment (CCME); Environment Canada; the International Institute for Sustainable Development (IISD); the International Development Research Centre (IDRC); and the National Round Table on the Environment and the Economy (NRTEE). Representatives from over 40 sectors of Canadian society attended the meeting, including business associations, community organizations and indigenous peoples.

Each of the five 'sponsoring' organizations, acting as a working group, contributed C$ 50,000 to establish a secretariat and hire a research director. Two subcommittees (Documentation and Information, and Vision and Process) assumed responsibility, respectively, to analyse Canadian responses to Rio, and to draft a concept paper on sustainability planning. The NRTEE facilitated and chaired the process and provided the secretariat. Most of the tasks were undertaken by volunteers and committees which met monthly. There were those who wanted to 'develop strategic plans' and others who wanted to 'do specific projects'. It was therefore decided to do both.

A progress report and recommendations were presented to a second national stakeholders' meeting in June 1993. At the third assembly in December 1993, the NRTEE was asked to assume a larger management role for the next phase of the project, rather than merely acting as a facilitator, and to move towards preparing a draft strategy. The NRTEE worked closely with a volunteer working group to develop, revise and review a strategy document. A draft was tabled at the fourth assembly in November 1994, entitled Canadian Choices for Transitions to Sustainability. *Minor changes were suggested and the document was endorsed. A revised document was published in January 1995. The NRTEE then organized a series of about 12 meetings across the country to determine how useful such a document might be in engaging various constituencies in discussions about sustainability. A final revised draft, based on the feedback received, was published in June 1995. The working group, which had been reconstituted in early 1995, in addition to completing the strategy document, developed a work plan involving, among other things, the compilation of a directory of sustainability tool kits for communities. 'Sustainable livelihoods' was selected as a focus for working group activities with a forum on this subject planned for March 1996.*

Principles of the Projet de société:

- *The process was designed to be transparent, inclusive and accountable.*
- *Each partner and sector was encouraged to identify and take responsibility for its own contribution to sustainability.*

- *Dialogue and cooperation among sectors and communities were key elements of problem-solving.*
- *A shared vision and agreement on key policy, institutional and individual changes were seen as necessary for the transition to sustainability.*
- *It was stressed that strategy and action must be linked, and must build on previous and ongoing initiatives.*
- *Canada's practice of sustainable development and its contribution to global sustainability should be exemplary.*

Projet de société (1993, 1994, 1995b)

Box 5.3 THE ROUND-TABLE EXPERIENCE IN CANADA

The various approaches to stakeholder participation in Canada include round tables on the environment and economy (of which Canada now has hundreds at national, provincial and local levels), multi-stakeholder task groups (such as the Climate Change Task Group and the Task Force on Economic Instruments and Disincentives to Sound Environmental Practices), and commissions, councils and collaboratives (such as the Economic Instruments Collaborative). These all attempt to bring together a broad range of competing interests to work on solutions; and they usually rely on consensus for decision-making and a neutral chair or facilitator.

The relationship to the policy process has been diverse. It has been used to develop broad strategies, to implement or monitor those strategies, to prepare principles or action plans which may then be 'self-implemented', to prepare policy options for government (for temporary or permanent issues), or to carry out public consultation phases in the development of public policy.

In themselves, almost all these elements form but a small part of the policy process, despite so much recent focus on them. However, Ronald Doering, the former executive director of the NRTEE, has noted that they have been described as 'innovative institutional adaptations that will play an increasingly important role in future years as we reinvent government by trying to improve our ability to engage citizens more deliberatively in policy choices'. Doering quotes a political scientist in claiming that 'the institutionalization of multi-stakeholder forums is the most significant innovation in the Canadian policy process in the past decade'. Yet he notes that they have also been vilified as 'superficial, mere window dressing, a waste of time, or a disguise for a vacuum by encouraging talk rather than action'. Some are suffering from 'consultation fatigue with the same few "elites" being consulted again and again. The corporate business sector displays less interest in these processes while ... NGOs have a declining capacity to participate' (Doering, 1993).

A few clear lessons and dilemmas come across in the NRTEE review:

(1) When designing the multi-stakeholder process, it is important to distinguish consultation from consensus. The former meets the needs of the initiating party, but the latter should be participant-driven, which requires a neutral facilitator. The role of a multi-stakeholder process is different in each case, but 'many of the frustrations

of past efforts have resulted from a lack of clarity on this ... or from an attempt to blend the two approaches. They do not blend easily. ... You can't have the buy-in and other advantages of a consensus process until you're willing to ... allow the participants to design and manage the process'. Specific examples are not, however, given.

(2) Neutral facilitation is needed to achieve round-table objectives, as people with very different value systems and even different vocabularies naturally find it hard to agree.

(3) The involvement of NGOs is essential, but many cannot afford to participate, especially to get involved in research and go beyond mere attendance. Yet government funding for NGOs compromises their independence and means that some NGOs are fully taken up with government-driven agendas. In other words, round tables could be seen as a way for governments neatly to 'contain' participation to a limited part of the whole policy process, and indeed to co-opt some groups. This has been a real problem in some round tables.

(4) In many circumstances, round-table approaches are inappropriate, either because of the subject matter, or because key stakeholders lack the necessary time or commitment. In particular, firm political commitment to act on possible outcomes is needed initially. Round-table processes are 'still in the development stage, and it is wrong to see them as a mature phase of the policy process.'

Yet, as current institutions are not coping well with the transition to sustainability, in part because of their jurisdictional fragmentation, round tables are worth pursuing because they force governments to 'take more seriously what they call the horizontality problem', i e cross-departmental cooperation.

Doering assesses that multi-stakeholder processes 'have been important experiments in policy-making and public administration. Their role is essentially transitional and catalytic; they support rather than replace elected bodies. With all their flaws, and while still generally marginal to core policy making, Canadian round tables are common-sense partnerships'.

Another commentator addresses the political aspects. He acknowledges that multi-stakeholder processes have helped environment, consumer and Aboriginal interests to be better represented in the 'policy marketplace'. But these processes may result in 'politically compelling consensus which constrains the ability of elected politicians to make decisions'. In other words, 'bargaining' through this marketplace is replacing the search for the common good. The better bargainers get the best deal, or, perhaps, 'organized interests bargain among themselves, cut up the pie and invite elected representatives to serve the helpings'. In effect, 'the utility of multi-stakeholder exercises should reflect both how and how well they assist elected representatives in their core task — searching for and defining the common good, and incorporating it in public policy'. A further commentator suggests that it is essential to have a neutral forum such as NRTEE; no one stakeholder could bring together the right group without raising suspicions.

The NRTEE has also examined the issues of representativeness, governance and democracy. The kinds of dilemmas raised include the notion of the flourishing of a stakeholder elite at the expense of the broader public's involvement in decision-making; stakeholder representation (stakeholders should be able to state who they are and who and what they represent); the need to make participation more transparent and involve more than an elite — and a broader network of stakeholders would reduce the burden on the 'over-consulted'. One commentator suggests that 'multi-

stakeholder processes mask significant imbalances in power over resources and considerable ... differences in influence on government among the participants'.

Further dilemmas present themselves when it comes to implementing round-table agreements and action plans. Dana Silk proposes the notion of 'sustainability mediators' — individuals in the various institutions whose job it is to liaise with other institutions and work on further consensus and joint management. He sees such people as specialists in working across different sectors, recognizing that this is a special skill which not all people have.

This approach has been used in some NCSs and NEAPs. Under the Pakistan NCS, for example, environment contact officials are appointed in key government agencies. However, it is not known whether the Pakistan government selected these contact officials on the basis of their aptitude.

The main benefits of the round-table approach tend to be in the various forms of consensus achieved. While fully recognizing that its approach is yet immature, the NRTEE issued, in 1993, a set of consensus principles which have been very widely distributed across Canada and many other countries. In brief, these principles (NRTEE/ParticipACTION, 1994) are:

1. *Purpose-driven (people need a reason for participation)*
2. *Inclusive, not exclusive (as long as parties have a significant interest)*
3. *Voluntary participation*
4. *Self-design (the parties design the process)*
5. *Flexibility*
6. *Equal opportunity (in access to information and participation)*
7. *Respect for diverse interests (and different values and knowledge)*
8. *Accountability (to parties both within and outside the process)*
9. *Time limits (realistic deadlines)*
10. *Commitment to implementation and monitoring*

(Projet de société, 1995a)

meetings, writing letters and preparing briefs. Environment Canada officials have commented that the aim was to strike a balance between leadership of the process and public participation. But the consultations appear to have satisfied very few. According to Toner (1994):

> *This was partly because everyone knew they were an 'add-on' to the process, as the original time limits imposed by [environment minister] Bouchard simply did not allow for consultation. Thus, many stakeholder representatives were already suspicious when the consultations began, and some were even more upset when they discovered that the document developed for the consultation process was not the actual policy document that had been presented to Cabinet in January [1990], but rather a discussion paper that asked a series of vague, general questions and did not contain any financial detail. This was hardly surprising since Cabinet had not yet decided on the total dollar figure for the program as a whole, nor had they approved any details of the plan. The lack of specificity in the content of the discussion*

Box 5.4 WHY CONSULTATION BECAME A LATE FEATURE OF THE CANADIAN GREEN PLAN

During its development, the Green Plan was linked to the budget process and therefore became subject to Cabinet 'secrecy'. Departments came forward with scores of programme proposals and a 'wish list' emerged with a potential nominal cost of C$ 10 billion. This figure was 'leaked' to the media and raised expectations.

To control further leaks, DoE [Environment Canada] officials restricted the access of other departments to the [Green Plan] document, and this upset officials from some departments. Charges of secrecy from within the system were amplified by charges from both businesses and environmental groups that they were not being adequately consulted. There was an emerging tradition of involving both industry and environmental groups in DoE policy development, which had been built up around the Canadian Environmental Protection Act process and other recent initiatives. The green planners who were new to the Department were not part of that tradition, and both industry and environmental groups complained that the briefings they were given on the Green Plan did not constitute consultation.

These charges of secrecy were inflamed by suspicions generated by the energy sector, because several key officials who were involved in drafting the Green Plan had worked at Energy, Mines and Resources during the early 1980s when the Liberal government introduced the interventionist National Energy Program. The oil and gas sector was concerned that a carbon tax might be introduced to encourage reductions in gas emissions that contribute to the greenhouse effect. As a party, the Conservatives had fought long and hard against the [programme], and rescinding it was one of their first initiatives in office. Linking [it to] the Green Plan . . . crystallized the attention of both Tory backbenchers and cabinet ministers and raised red flags. When the draft Green Plan was taken back to the Priorities and Planning Committee for approval in January 1990, the Cabinet did not decide definitively on the plan but instructed Bouchard [the environment minister] to develop a process to consult with the public.

(Toner, 1994)

document meant that consultation sessions were not directed either to choosing priorities or exploring detailed program options. Many charged that the real document was locked in a desk back in Ottawa, while the consultations were just window-dressing. In spite of these misgivings, several thousand people participated in the consultations.

Despite these problems, it needs to be remembered that the Green Plan was developed before UNCED (1992) when the notion of participatory policy development really took hold in Canada and in other countries.

In developing the Netherlands' second *National Environmental Policy Plan* (NEPP2)

(VROM, 1993a), a network of some 600 individuals from across government departments, research institutions, target groups, NGOs and local government was involved in writing and/or participating in meetings, conferences and round tables. But the NEPP2 was still very much a 'consultative' process (the first NEPP, 1991, was an internal government undertaking with no real involvement of the public, industry or NGOs). The aim was to focus on persuading target groups to engage in discussing what changes they should make and to participate in monitoring implementation.

The trend in the Netherlands is to move away from direct top-down (command and control) economic and environmental management, to more socially negotiated and participatory instruments, such as voluntary covenants with target groups that set out monitoring and evaluation systems. Covenants, agreed under NEPP2, were always the result of a participatory process with the target groups. Such negotiation and public discussion are part of Dutch culture. There is a long tradition of government cooperation with industry and local authorities. Covenants, which started essentially as 'gentlemen's agreements' with a highly uncertain status and degree of enforceability, are now generally standardized and formalized with regard to procedure and content. The government sees them as a way of expressing joint responsibility (see Box 14.3). About 100 covenants have been concluded between the government and the private sector in recent years. It is still too early to judge the real success of the voluntary agreements struck under NEPP2. A recent *OECD Environmental Performance Review of the Netherlands* concludes that:

> *progress has been made in the implementation of such agreements: they are stimulating cost-effective actions, and serve as more or less binding substitutes for regulation in a number of areas. However, should the environment cease to be a major public concern, it is not certain that these instruments would lead to substantial results. They must be used in association with other instruments and with mechanisms of accountability.*
>
> (OECD, 1995b)

NGOs had considerable indirect influence on the NEPP and NEPP2 via their activities in promoting public campaigns. They also had direct influence through participating in working groups and round-table meetings. The Ministry of Housing, Spatial Planning and the Environment (VROM) has stated that it would like to see NGOs increasingly change their attitude from the traditional one of confrontation to become partners and participants in the NEPP process.

In the UK, interest groups were kept more at a distance, although some were shown drafts of some chapters of the *UK Strategy for Sustainable Development* (UKDoE, 1993). In Norway, they were not shown strategy drafts, but they were asked for their views on the Brundtland report, and the findings were fed into the green planning process by government bodies.

France's *Plan national pour l'Environnement* (PNE) (French MoE, 1990a), while still partly consultative, was less so than others described above. It was developed mainly through formal consultations and numerous meetings with, and contributions from, all relevant government ministries, agencies, regional delegations on architecture and on the environment, and the Federation of Nature Parks. The plan thus evolved mainly as an interdepartmentally-negotiated text. There were two principal writers in the

Ministry of Environment who coordinated a 12-person PNE preparation team (mostly staff of the Ministry of Environment). However, informal meetings and discussions were held with industries, businesses and NGO associations. Written contributions from NGOs on their ideas about their role in the decision-making process were requested and received for inclusion in the PNE; and most of these ideas were taken into account. There were also some informal discussions with representatives of various cities and with a range of politicians.

About midway in the development of the PNE, a public meeting was organized for some 300 invited people. It was attended by the president and Prime Minister, and by all ministers who were asked by the PM to present their ministerial responsibilities, policies and actions concerning the environment. This was the first occasion in France that such a meeting on the environment had been held. An open meeting was later held in seven of France's 22 regions, at which the draft PNE was presented for discussion.

There was no formal process of consultation in preparing the European Union's Fifth Environmental Action Programme, which was drawn up before the current 'era of transparency'. Nevertheless, informal consultations took place between the European Commission and NGOs, but these openings were not followed up by NGOs and the European Commission was criticized afterwards for a lack of consultation. The Fifth Environmental Action Programme 'team' travelled to each of the member states to elicit views and priorities. Once the programme had been approved, the European Commission funded a series of open meetings in each member state, hosted by the European Environmental Bureau (a Brussels-based organization which represents environmental NGOs from the 15 member states) to discuss its implications.

In a similar way, the drafting of Latvia's 1995 *National Environmental Policy Plan* (NEPPL) involved members of a wide range of government departments and agencies, academic institutions, and some NGOs and private consulting companies with environmental expertise. NGOs were invited to comment on drafts and participate in seminars.

In industrialized countries, NGOs have been involved, often fairly extensively, in providing information for national strategies. Hill (1992) compared the national environmental plans of Canada, France, the Netherlands, Norway and the UK. All of these plans were published as government documents, and most of the implementation is supposed to be the responsibility of government and the private sector. In reality, NGOs were involved only in data gathering and consultation.

The earliest United Kingdom initiative, the Conservation and Development Programme, was released in 1983 in response to the World Conservation Strategy. It was dominated by NGO input, which had been made with inadequate involvement of government at the highest level. This meant that the government felt unable to deal with the large agenda sprung upon it with the publication of the programme, which consequently led to little real action. By contrast, in the UK initiative *This Common Inheritance* (HMSO, 1990), the government was dominant and there was no formal consultation process. The views of some NGOs, including those of environmental pressure groups and trade associations, were canvassed only informally, and this led to considerable criticism by a large segment of the public. The latest UK initiative, *Sustainable Development: The UK Strategy* (HMSO, 1994a), now includes a recognition of what NGOs can do for sustainable development; yet the NGO input into the (government-led) process took the form of consultation rather than participation (Dalal-Clayton et al, 1994).

In Latvia, some displeasure has been expressed by certain NGOs, particularly with respect to the central and eastern European Environmental Action Programme seminar. There are still relatively few active and experienced NGOs existing in Latvia and more recently the existing NGOs have suffered a decline in popularity. During the late 1980s, when the Soviet regime was crumbling, the environmental movement gained considerable popularity, and many of the environmental campaigns clearly had political implications. Between 1990 and 1993, however, many NGOs lost public support, for political changes, with the emergence of a market economy and a decrease in the standard of living, undermined public interest in environmental issues.

In the USA, most of the work of the task forces set up by the President's Council on Sustainable Development (PCSD) was of a consultative nature. The Natural Resources Task Force held workshops and case studies on watersheds across the country, and members listened to local and regional stakeholders share their views on the importance of natural resources, watersheds and a sustainable future.

There was a reasonably good level of involvement of national-level NGOs, but minimal input from grass-roots NGOs. While the latter appeared to support the concept of the PCSD, they were concerned that this may result in 'unwinding' some of the hard-won regulatory mechanisms and legislation, and eliminating some environmental protection measures. They were also reluctant to become fully engaged because of the high demand in terms of cost and time.

Some PCSD task forces were able to support travel costs (but not the salaries or fees) of NGOs and grass-roots organizations. By comparison, industry could 'afford' to be involved. A strong effort to maintain a balance was made by the co-chairs and the executive director.

Predominantly Internal Government Processes

The development of Denmark's new *Nature and Environment Policy Plan* (DanMoE, 1995) typifies processes that are essentially internal to government. It was managed by the minister's secretariat in the Ministry of the Environment (MoE), but on behalf of the government as a whole. Consultation and negotiation took place only among government ministries and agencies. The need for the plan, the concept and its structure were discussed with all relevant ministries as a collaborative venture. Most ministries preferred the MoE to take the lead in writing sector chapters, which the others would then review and comment on. Some ministries made initial contributions. Draft sections were discussed and negotiated in numerous meetings.

The main drafting work was effected through a working group comprising representatives from the MoE, agencies under the MoE (Environmental Protection Agency, Nature and Forests Agency, Energy Agency), and environment and geology research institutions. This was carefully scrutinized at higher-level meetings within the ministry's various divisions until all the more difficult problems had been ironed out and any outstanding conflicts resolved. The final arbiters in any disputes that arose were the directors-general of the various agencies concerned, who formed a committee for this purpose.

In the initial stages, dialogue involved the various agencies of the different ministries. A seven person drafting/editorial team, comprising the best writers from different agencies, was established in the MoE for the final stages. This team engaged in intensive dialogue with the Minister of Environment's secretariat. It is intended to discuss the proposals contained in the published policy with NGOs, other organizations and industry, particularly in terms of future initiatives.

In Sweden, soon after UNCED, Agenda 21 and the Rio conventions were translated into Swedish. These documents were circulated by the Ministry of Environment to some 300 organizations in the country, including government departments, local authorities, NGOs and individuals, for review and comment. In Spring 1993, on the basis of a joint letter from the MoE and the Swedish Association of Local Authorities to all local authorities, they were asked to consider how UNCED agreements tallied with the present situation and policies in the country and to recommend what changes were needed. A summary of Agenda 21 was prepared and 25,000 copies were distributed nationally. About 250 organizations submitted written responses proposing actions and/or changes. The MoE analysed the public responses and prepared a draft outline of Bill 1993/4.111, *Towards Sustainable Development in Sweden* (SwedMoE, 1994). This was circulated to ministries for review and comment. Many rounds of negotiation followed between government departments, but there was no involvement of industry, NGOs or the public in drafting the bill itself. This was an internal government process. Similarly, development of *Enviro '93* — the action programme of the Swedish Environmental Protection Agency (SwedEPA, 1993) — was undertaken mainly as an internal process within the EPA. However, some of the underlying reports were developed, more or less, with the cooperation of sector agencies. For example, the work on the report on agriculture was led by the National Board of Agriculture. The same process has been followed by the US EPA in its Environmental Goals Project, although round tables and other consultative meetings have been held — but attendance has been by invitation only.

After decades of central planning, the governments of the countries of central and eastern Europe are not oriented towards people's participation. At the same time, the public are unaccustomed to 'participating' in government decision-making. A recent report (REC, 1995) on progress with NEAPs in the CEE countries notes that:

> *Open discussion of environmental goals and objectives of environmental action-oriented plans is not frequent in the CEE countries. The public is not involved in evaluating ways to strengthen environmental administrations and their role is often reduced to commenting on programmes already in the draft stage. The current investment policy in certain countries practically eliminates the public from monitoring the environmental impact of established facilities. There is still no mechanism for the public to participate in steering committees or to provide comments on the work of investors.*
>
> *The role of NGOs in monitoring environmental protection programmes is different in each country. It varies from having a very weak involvement to having a relatively strong influence on the environmental decision-making process. In regard to the NEAP process, their involvement has also taken the form of commenting on drafts of governmental programmes. Thus public participation still has a passive role and little impact on environmental protection programmes.*
>
> *There has been, as with priority setting, more progress with public involvement at the county and municipal level. Public engagement has been more influential in deciding goals and objectives of environmental protection.*

Thus, in Poland, for example, the public did not participate in either the development of the *National Environmental Policy* (PolMeP, 1990a) or the subsequent Implementation Plan (PolMeP, 1994a). It was formulated by a panel of Poland's leading environmental experts. During formulation of the Implementation Plan, some effort was made to communicate with the *gmina* (local administrative body — a cluster of villages) through questionnaires, but these were aimed at collecting factual information and not at providing options for national environmental policy.

Chapter 6 LINKS to OTHER STRATEGY and PLANNING PROCESSES

The development or implementation of some strategies has been influenced by, or they themselves have contributed to the development or implementation of, regional (supranational) strategies or initiatives.

Regional Strategies

Western Europe

Growing out of the original European Economic Community (EEC), the emergence in 1993 of the new and more ambitious 12-member state European Union (EU) is set to have far-reaching implications for environmental strategy-making across the region as attention shifts from adding an environmental dimension to the barrier-free internal market, to the more general pursuit of sustainable development.

Following the 1972 UN Conference on the Human Environment in Stockholm, the European Economic Community (EEC) launched the first of five multi-annual Environmental Action Programmes to define legislative priorities. By the early 1990s, the combined impact was considerable. More than 200 pieces of EEC environmental legislation had been adopted, and the UK's Secretary of State for the Environment, John Gummer, has estimated that 80 per cent of British environmental legislation is now decided collectively at the European level (Gummer, 1994). This move to Europeanize environmental policy and legislation has also proved to be highly popular with the public, who repeatedly cite European environmental activities as one of the most popular features of the integration process.

The signing of the Treaty of European Union at Maastricht in February 1992 shifted the emphasis of European environmental action by making the promotion of 'sustainable and non-inflationary growth, respecting the environment' (Article 2) an overriding task of the EU as a whole. It also tightened specific environmental provisions, notably by requiring environmental protection requirements to be 'integrated into the definition and implementation of other Community policies' and added the precautionary principle to the existing list of guiding themes for policy-making (such as the polluter pays and prevention principles).

Set against this extension of Community competence, the Maastricht Treaty also emphasized the principle of 'subsidiarity', which has two important implications: first, European action should take account of the diversity of the European Union, a diversity that increased further with the entry of Austria, Finland and Sweden as new member states in January 1995; and second, that European action should be taken only when environmental problems cannot be better solved at the local or national level.

But despite all these efforts, the Community's 1992 *State of the Environment* report (CEC, 1992b) showed a 'slow but relentless deterioration of the general state of the environment'. A new approach to strategy-making was clearly required and, in early 1992, the European Commission presented its proposals for a Fifth Environmental Action Programme. Entitled *Towards Sustainability*, the document marked a distinct break from the EEC's previous four environmental action plans (CEC, 1992a):

- it shifted the emphasis away from protecting the environment in favour of promoting sustainable development by integrating social, economic and environmental factors in decision-making;
- it took a strategic approach, setting long-term objectives for a number of priority issues, outlining indicative targets for 2000 and identifying a broad range of instruments for changing patterns of behaviour; and
- it focused on five key target sectors responsible for the bulk of environmental damage: agriculture, energy, industry, tourism and transport.

The Fifth EAP (see Chapter 21) was driven by the EU's own need to provide a strategic response to the 1987 Brundtland report and to detail the implications of the new Maastricht provisions. But it also reflected developments at the national and international levels. Nationally, the Netherlands' *National Environmental Policy Plan* (VROM, 1989) proved highly influential in setting out the consensus-based target group approach to strategy-making. Internationally, the preparations for UNCED provided a focus. Its design and follow up, however, highlight the EU's uneasy transition from an intergovernmental to a federal institution.

The original programme was drafted during November 1991 by the environment directorate general of the European Commission, which has the mandate for proposing legislation. It was then laid before the Council of Ministers representing the then 12 member states and the European Parliament, the directly elected body representing EU citizens. While the European Commission had engaged in some prior consultation, the Council of Ministers could not change the programme. It adopted a broad resolution in February 1993, highlighting its priority concerns, and approving the programme's 'general approach and strategy'. Meanwhile, the European Parliament could only pass an advisory resolution. According to one official from the UK Department of the Environment, this process meant that 'the Programme was essentially a European Commission statement on what it intended to do; it was not a binding legal document. The European Council of Ministers was not entirely in control of the Programme and this was reflected in its Resolution' (Plowman, 1993).

The implementation of the Fifth EAP over the past three years has, to some extent, tackled this central question of ownership. Three 'dialogue' groups have been set up to help with the implementation of the programme: the General Consultative Forum; the Environmental Policy Review Group; and an informal network on implementation and enforcement of legislation (IMPEL) (see Chapter 21, p232).

The European Commission has also taken some important steps to put its own house in order. Following the final approval of the Fifth EAP, the Commission approved a set of internal measures to integrate the environment throughout its operations into all other areas. Each of the European Commission's 23 Directorates-General dealing with different policy areas now has a high-level official to ensure that policy and legislative proposals take account of the environment and contribute to sustainability. The DGs must also carry out regular evaluations of their environmental performance and report on progress. The Environment Directorate has set up its own unit to coordinate these efforts and build up the necessary environmental management skills within the European Commission to make integration a reality.

In some areas, the results have been impressive. In December 1993, the European Commission launched its plan for economic recovery — the White Paper on *Growth, Competitiveness and Employment* (CEC, 1993a). This is perhaps the first macro-economic strategy to incorporate the goal of sustainable development as an economic imperative. New measures have also been agreed to incorporate the environmental dimension into the planning and disbursement of the structural and cohesion funds for poor areas within the European Union. But the Commission itself recognizes that 'sustainable development essentially continues to be seen as the business of those who deal with the environment' (CEC, 1994a). The Common Agricultural Policy, which still consumes the largest proportion of the EU budget, remains largely unchanged by sustainability factors.

The issue of linking up strategic efforts at the EU level with national and local initiatives has proved highly sensitive and is discussed in detail in Chapter 21 (see sections below on links to national planning and on links to local strategies (p229), and within member states and others (p232)). The European Commission was aware of national planning initiatives, particularly in France, the Netherlands and the UK. It was particularly influenced by the Dutch *National Environmental Policy Plan* (NEPP) (VROM, 1989) and adopted the target group approach which it had pioneered. The Commission never sought to posit the Fifth Environmental Action Programme as an 'overarching framework' for coordinating European and national-level strategy processes, although some countries (notably Greece, Italy and Portugal) used it as a model. Other countries were sceptical about the need for a European strategy. Some European NGOs, however, have continued to press for such a European framework for national strategies.

The EU is now moving into a federal environmental age, where the European level is not just a legal arena for law-making, but a forum for learning and sharing best practice. Looking ahead, the EU will need to resolve whether the broad strategic approach taken by the Fifth Environmental Action Programme should be transformed into a tighter planning process, with binding targets for achievement. It will also need to find ways through the inherent complexity of sharing the responsibility for sustainable development with the multiple tiers of government and the almost infinite range of economic, social and civil partners across the EU.

These tensions between the European, national and local levels of strategy-making

Box 6.1 RECENT EUROPEAN ENVIRONMENTAL INITIATIVES

The Bergen conference (May 1990) on sustainable development in the UN Economic Commission for Europe (UNECE) region emphasized the need to improve reporting on the state of the environment (many European countries now publish such reports regularly — sometimes as part of strategy processes). Prompted by the changes occurring in central and eastern Europe following the break-up of the former Soviet Union, a joint meeting of environment ministers from this region, the European Union and the European Free Trade Area (EFTA), was held in Dublin in June 1990. This meeting produced the idea of holding regular European ministerial conferences to address environmental matters.

The first of these 'pan-European' conferences took place in Dobris Castle in former Czechoslovakia in June 1991. Here, there were calls for a report assessing Europe's environment. The subsequent report, Europe's Environment: The Dobris Assessment *(Stanners and Bourdeau, 1995), covers the state of the environment in nearly 50 states. It was prepared by an EC task force for the new European Environment Agency in cooperation with the UNECE, UNEP, OECD, the Council for Europe, WHO, IUCN and Eurostat, together with the individual countries of Europe. The report is intended to 'provide an objective basis for planners and developers involved in policy-making and programming in environment and sectorial fields' (EEA, 1994). This report has not yet been able to influence the development or revision of any strategies.*

The second ministerial conference was held in Lucerne and endorsed the Environmental Action Plan for Central and Eastern Europe. A third conference was held in October 1995 in Sofia and endorsed an Environmental Programme for Europe (EPE) (Box 6.2).

will be accentuated as the EU expands further to include central and eastern European (CEE) countries over the coming decade. In June 1993, at the Copenhagen summit, EU heads of government resolved that the CEE countries could become members of the EU 'as soon as they are able to fulfil the necessary conditions'. To implement this agreement, the EU has approved wide-ranging European agreements with Bulgaria, the Czech Republic, Estonia, Hungary, Latvia, Lithuania, Poland, Romania and Slovakia, and is preparing these associated countries for membership through a preaccession strategy endorsed at the December 1994 EU summit in Essen.

The preparations for EU membership will have an important impact on sustainable development strategy-making, in both eastern and western Europe. Both sides are now exchanging information on environmental policy and strategies for sustainable development and are evaluating priorities for the eventual convergence of central and eastern European countries' environmental law towards EU norms. The longer-term implications of an EU stretching potentially from the Atlantic Ocean to the Ural Mountains will be profound, centring on the EU's continued ability to deepen its drive towards sustainable development, while widening its membership to an ever more diverse range of countries.

Central and Eastern Europe

Agenda 21 outlined actions needed to halt and reverse environmental degradation and bring about sustainable development in all countries, and stressed that the responsibility for environmental protection had to be shared by all countries. These concerns have also been the subject of a series of European initiatives (see Box 6.1). One of these initiatives was the comprehensive '*Environmental Action Programme* (EAP) *for Central and Eastern Europe*' (UNECE, 1995), which spelled out a process to equalize environmental conditions in the east and west, with an emphasis on the urban environment (see Box 6.2). The EAP is not a final document with specified objectives that have to be precisely followed and implemented by CEE countries. It is rather an instrument or methodology that countries can use to draft viable environmental protection plans. Progress made by CEE countries in adopting national environmental action plans has recently been reviewed by the Regional Environment Centre for Central and Eastern Europe (REC, 1995) (see Table 2.2).

The EAP document was used in Latvia as a 'handbook' for the development of the *National Environmental Policy Plan*, 1995. However, only some aspects of the EAP were drawn upon, other aspects being more relevant to other CEE countries, for example, the emphasis on human health and air pollution. In Poland, the *National Environmental Policy* (NEP) (PolMEP, 1990a) was developed well before the EAP for Central and Eastern Europe, but the NEP Implementation Plan to Year 2000 (PolMEP, 1994a) was able to take account of ideas in the EAP.

But national initiatives in eastern Europe have also spawned regional responses. In 1991, the Polish government launched the '*Green Lungs of Poland*' initiative, aimed at combining nature protection with economic development in northeastern Poland (NFEP, 1991). This is an area very rich in biodiversity and the strategy was developed with participation of the *voivodships* (provincial authorities). It has already resulted in the abandonment of plans to open a mine. More recently, this initiative has been extended as the '*Green Lungs of Europe*' initiative encompassing a much larger area — Belarus, Estonia, Latvia, Lithuania, Poland, Russia and Ukraine — and the 'Green Lungs of Europe' international agreement was signed in Warsaw in February 1993 (ISD, 1993). It aims to 'protect regions richest in biodiversity as well as to increase the environment's capacity to accumulate carbon in biomass'.

North America

The development of Canada's *Green Plan* (Government of Canada, 1990a) involved interfacing with and taking account of several regional initiatives. The Circumpolar Conservation Strategy, which was originally called the Finnish Initiative and was developed by agreement between eight nations (it later led to the Circumpolar Arctic Environment Protection Strategy 1991), strongly influenced the drafting of Green Plan programmes to address issues in the Arctic. In fact, Canada's implementation of this strategy was funded under the Green Plan (C$ 100 million). Similarly, while the Canada–USA Great Lakes Water Quality Agreements (GLWQA) preceded the Green Plan, the latter became a vehicle for expanding and extending actions to meet the commitments under the GLWQA. The Green Plan was also the basis for Canada's negotiations at UNCED (although the process to develop Canada's UNCED national

Box 6.2 THE ENVIRONMENTAL ACTION PROGRAMME FOR CENTRAL AND EASTERN EUROPE, AND ENVIRONMENT PROGRAMME FOR EUROPE

In April 1993, European environment ministers met in Lucerne and endorsed the Environmental Action Programme *(EAP)* for Central and Eastern Europe. *Between US$ 30 and $50 million in grant funds was pledged to support better project identification and preparation, as well as numerous small investments to generate environmental benefits.*

The EAP was prepared by the World Bank, with the OECD as part of a task force chaired by the European Commission. It proposed to concentrate on the following activities:

- *Environmental planning to include setting priorities, identifying tools and measures of effective environmental management and involving the public in environmental decision-making;*
- *Institutional strengthening to increase efficiency in environmental administrations and to improve environmental monitoring, control and enforcement of environmental regulations; and*
- *Environmental investments that address immediate local and regional problems and identify long-term sustainable solutions.*

Using practical examples, the EAP indicates how economic and sectoral policies and investments can best contribute to environmental improvement. It adopts the basic premise that, due to scarce financial resources, human health impacts must be the primary criterion in setting environmental priorities. The focus, therefore, is mainly on pollution problems that affect both cities and the surrounding countryside and are common to the countries in the former Soviet bloc.

The programme also encourages consensus among central and eastern European countries and donor agencies on environmental priorities. It promotes a mix of policy, investment and institutional actions such as cutting air emissions from specific types of industrial plants, reducing particulate and sulphur dioxide emissions in urban areas (especially linked to the use of coal in the household and service sectors), launching low-cost, high-gain programmes such as energy efficiency and environmental audits in the industrial sectors responsible for the most pollution, protecting groundwater from wastewater discharges and hazardous wastes, and undertaking municipal wastewater investments for improving ambient water quality at low cost.

In the first year, a range of actions to implement the programme were initiated, including translation of the programme into 19 eastern European languages; preparation by the Kyrgyz Republic of a national environmental strategy organized around the themes of the regional action plan; and a training seminar in Bulgaria for 38 officials from ministries in various countries.

The Environmental Programme for Europe *(EPE) was endorsed by the third pan-European ministerial conference in Sofia, Bulgaria in October 1995. The EPE sets out long-term environmental priorities at the pan-European level (i e much wider than the EAP for Central and Eastern for Europe) and addresses many of the issues raised by the Dobris Assessment (see Box 6.1). It comprises a number of key policy actions, other recommendations, and a separate background document giving further information on policy in the various areas covered. The EPE emphasizes a number of common cross-sectoral issues shared by all countries of the region. It also seeks to promote the work of the European Environment Agency (EEA). The EPE contains a wide range of recommendations, recognizes the importance of public participation 'at all levels of environmental policy-making', seeks to strengthen capacity-building and environmental education systems, recommends more attention to compliance monitoring and enforcement procedures for environmental legislation, and promotes the integration of environmental considerations in all key sectors*

Specific recommendations are made for:

- *cleaner production and efficient use of energy and materials, sustainable consumption patterns (improved energy efficiency and renewable energy; environmentally sound technology, and waste management);*
- *sustainable consumption patterns (market-based instruments, eco-labelling, environmentally friendly transport, recreational activities, urban stress including human health aspects);*
- *sustainable management of natural resources (soil protection, integrated and sustainable water management — in particular in transboundary waters, integrated coastal zone management and the protection of the marine environment, and spatial development);*
- *biological and landscape diversity; and*
- *sustainable agriculture, forestry and fisheries.*

Sources: World Bank (1994); UNECE (1995)

report was considerably more open than that for the Green Plan itself). Subsequently, the Green Plan itself formed the basis of Canada's position in its discussions with the USA and Mexico on the North American Agreement on Environmental Cooperation (final draft, 13 September 1993) — negotiated as part of the North American Free Trade Agreement (NAFTA).[1] The agreement on environmental cooperation had ten main objectives (see Box 6.3), and Environment Canada staff have commented that it would probably not have been concluded without the Green Plan being in place. This was mainly because, in the Green Plan, the government undertook a commitment to implement the Canadian Environmental Assessment Act, which stipulated that environmental factors had to be considered in the assessment of all cabinet submissions, and this meant that NAFTA was subjected to an environmental assessment.

It is unclear to what extent, if any, the agreement influenced the work undertaken by Canada's *Projet de société*. But in the USA, it had little apparent influence on the work of the President's Council on Sustainable Development to create a sustainable development action strategy, or on the Environmental Protection Agency's Environmental Goals Project.

1. NAFTA, which took effect in January 1994, outlined tariff cuts and the elimination of trade barriers between the USA, Canada and Mexico.

Box 6.3 OBJECTIVES OF THE NORTH AMERICAN AGREEMENT ON ENVIRONMENTAL COOPERATION

- *Foster the protection and improvement of the environment in the territories of the parties [USA, Canada, Mexico] for the well-being of present and future generations.*
- *Promote sustainable development based on cooperation and mutually supportive environmental and economic policies.*
- *Increase cooperation between the parties to better conserve, protect and enhance the environment, including wild flora and fauna.*
- *Support the environmental goals and objectives of the NAFTA.*
- *Avoid creating trade distortions or new trade barriers.*
- *Strengthen cooperation on the development and improvement of environmental laws, regulations, procedures, policies and practices.*
- *Enhance compliance with, and enforcement of, environmental laws and regulations.*
- *Promote transparency and public participation in the development of environmental laws, regulations and policies.*
- *Promote economically efficient and effective environmental measures.*
- *Promote pollution prevention policies and practices.*

Source: Canada/USA/Mexico, 1993.

Links to Budget Processes

While all strategy processes have financial implications if and when their policy and institutional recommendations are implemented, only a few have been directly linked to government budgetary processes.

The development of Canada's *Green Plan* (Government of Canada, 1990a) was based on a pragmatic approach to environmental management, and the fiscal aspects dominated. A relatively large proportion of 'new' government money was potentially available to implement the Green Plan and this was at a time of more general financial restraint and retrenchment. Environment Canada thus found itself evolving into a major department (with a significant budget). The Minister for the Environment (Lucien Bouchard) succeeded in integrating the Green Plan with the federal budget process. Toner (1994) points out that:

> *The 'linking' also implied that the Green Plan would be an expensive budgetary item and would have to be treated like a budget document with all the corresponding conventions of budget secrecy. In addition, it created a horrendously brief deadline of four months for the creation of a program that was system-wide in scope and would require extensive interdepartmental collaboration.*

Thus, the Green Plan had built-in targets and schedules — as a mechanism for public accountability. Politicians were able to be held publicly accountable. During discussions and negotiations between government departments over technical documents, there were parallel financial negotiations on budgetary allocations. A budget of C$ 3 billion was established for a five-year Green Plan. This was subsequently 'diluted': first the period was extended (in the 1991 budget) to six years; then reduced to $2.5 billion over two successive subsequent budgets (1992 and 1993). In practice, 60 per cent of the Green Plan programme and fiscal content falls under the jurisdiction of departments other than Environment Canada.

Similarly, in the Netherlands, the second *National Environmental Policy Plan*, NEPP2 (VROM, 1993a) presents environmental planning as a continuing process, which, among other things, includes the provision of an annual rolling three-year environmental programme of measures and actions alongside the parliamentary budget.

New Zealand's draft *Environment 2010 Strategy* (NZMfE, 1994a, p53) points out that the

> *annual government budget cycle provides an opportunity to consider the environment strategy in the broader context of the government's overall strategy and priorities.*
>
> *The annual budget cycle now includes a strategic phase for establishing the government's strategic priorities in the short, medium and long term. These strategic priorities are 'bedded in' through budget appropriations, Purchase agreements and strategic and key result areas in chief executive's performance agreements.*
>
> *Chief executives of government departments which affect the environment will be asked to take into account in their annual planning the goals of this strategy that are relevant to their responsibilities.*

The linkage between the annual budget and planning cycle and the planned four-yearly review of the *Environment 2010 Strategy* is shown in Figure 15.1. In presenting proposals for the follow-up to the report of the World Commission on Environment and Development (WCED, 1987), the Norwegian *Report to the Storting No 46* (NorMoE, 1989) introduced the idea of a separate 'green budget' (see Box 6.4).

Not surprisingly, the US Environmental Protection Agency's Environmental Goals Project (USEPA, 1994) is influencing the agency's budget planning process — it is using the current environmental goal areas in developing a goals-based budget for the 1997 financial year.

Links to National Planning

From documentation and discussions with key actors, it is difficult to ascertain the extent to which green planning initiatives and sustainable development strategies have influenced, or even been linked with, mainstream national planning. The latter tends to be the responsibility of, or is largely driven by, finance ministries. While some strategies have clearly been linked to budgetary processes within government and have therefore influenced decision-making and action planning by government and individual ministries, others appear to have had little connection at all with national planning.

Denmark's new *Nature and Environment Policy* (DanMoE, 1995) was developed to initiate a process of strategic environmental planning in the country. The intention was to build on several existing elements, namely regular ministerial-sector strategies, a range of periodic policy and state of the environment reports, and other policy-oriented

Box 6.4 NORWAY'S GREEN BUDGET

Each year, the Norwegian government's overall budget is presented to Parliament in early October. The budgets for all ministries are consolidated within one document. The idea of a separate 'green budget' — drawing together and making visible what budget flows in all ministries are directed to the environment — was first discussed within the Prime Minister's Office in August 1987. Cabinet approved the approach and there was detailed discussion with individual ministries. The Ministry of Finance wrote to all ministries providing guidance on preparing such a green budget within a framework for expenditure increases/decreases.

The 'green budget' is a published extract of the main budget but is amplified and illustrated as a separate document, with the aim of reinforcing the environmental responsibility of each ministry. The portions of ministry budgets that serve the environment are divided into three parts: money directed specifically towards environmental improvement, money spent with multiple aims, and funds for other purposes but which have incidentally positive environmental effects. However, there is no measure of the environmental efficiency of such procedures.

The green budget was first introduced as part of the 1989 budget, presented to Parliament in October 1988. The aim was to make this green budget document a 'steering tool' for Parliament concerning the environment. Unfortunately, Parliament has not really used it.

papers. The Ministry of Environment (MoE) aimed actively to promote consolidation and move further towards an integrated approach in which individual ministries would retain their responsibilities and answer to Parliament on how they were fulfilling environmental objectives in their areas of concern. Though coordinated by the MoE, the new policy plan is an overall government report. And though it contains no financial commitments, each line/sector ministry will be required to revise its own plans in accordance with government assurances set out in the new policy plan.

In France, the objectives of the *Plan national pour l'Environnement* (PNE) (French MoE, 1990a) were discussed, and incorporated, within the eleventh five-year development plan. France now intends to diversify the planning system by establishing territorial development directives to be drafted with the major local authorities and then approved by the state. They will aim to enable it to ensure respect, 'at the proper scale', of the necessary balance between development prospects and the protection of natural areas, sites and landscapes (French MoE, 1995).

The first Dutch *National Environmental Policy Plan* (NEPP) (VROM, 1989) was developed as an 'umbrella plan' paralleled by associated four-yearly sector reports (for example, on energy conservation, the state of water works, and the strategic plan for the development of infrastructure), each of which contained a 'heavy element' of environmental policies. The NEPP was strongly linked to key areas of sector planning and was formally agreed (signed) by five key ministries (Agriculture, Economic Affairs, Energy, Transportation, and Water Works). The subsequent NEPP2 (VROM, 1993a) sets out the conditions for the integration of sector policies of individual ministries with environmental objectives. Consideration is currently being given to further integrating physical planning in the Netherlands in the NEPP3 to be published in 1997.

New Zealand's *Resource Management Act* (RMA), 1991, is a major piece of reforming legislation concerned with the sustainable management of natural and physical resources. As such, it is related integrally to functions concerning national planning and decision-making (see Box 6.5), although it is not itself a planning instrument. Grundy (1993) comments:

> *The reform process [repeal of the Town and Country Planning Act (1977) and enactment of the Resource Management Act (1991)] was not only a rationalization of existing, admittedly often overlapping and contradictory, resource legislation, but also a deliberate move to limit the role of statutory planning in resource allocation decision-making. The wider socioeconomic objectives of the former legislation were viewed as unnecessary and undesirable interventions in the functioning of the market allocation mechanism and were removed.*

While Norway's *Report to the Storting No 46* (NorMoE, 1989) introduced the 'green budget' idea (see Box 6.4), the process of developing the report, at its inception, was poorly linked to national (particularly economic) planning. This is dealt with within the long-term programme produced every third year by the Ministry of Finance, which analyses all sectors, the economy and employment. However, the report was subsequently 'adapted' to the long-term programme being prepared at the time and 'worked around' elements that were being included in it — not the other way round.

Sub-National Strategies

Among the strategies covered in this study, few have been linked formally with (ie built directly on, or led directly to the development of) sub-national strategies at the provincial, territorial or state level. Nevertheless, in some countries, particularly those with a federal structure, such sub-national strategies are common, notably in Australia, Canada and the USA.

In Australia, each state or territory has independently developed or is preparing 'subsidiary' strategies to the *National Strategy for Ecologically Sustainable Development* (NSESD) (CoA, 1992a) — some environmental, some dealing with sustainable development, some both. Since all states and territories have endorsed the NSESD, they take account of it. For example, South Australia has developed a state conservation strategy (its equivalent to an ESD strategy) and a water plan.

Canada has a rich variety of sub-national strategies developed by different jurisdictions (see Box 6.6). While the *Green Plan* (Government of Canada, 1990a) drew from and stimulated some of these initiatives, there was no formal relationship between them and the independent *Projet de société*. It was an aim of the National Round Table on the Environment and the Economy (which facilitated the *Projet de société*) to develop a way of dealing with national issues that these and other sub-national strategies have been unable to handle, for example elaborating key issues and gaps in these other strategies, or those they have dealt with but which the *Projet de société* can help to resolve.

Box 6.5 NEW ZEALAND'S RESOURCE MANAGEMENT ACT (1991) AND PLANNING

The RMA is concerned with dealing with three conceptually separate but related things:

- *allocating access to commonly managed (public) resources such as water, coastal space (including coastal and estuarine water), the surfaces of lakes and rivers, riverbeds, the seabed, and geothermal energy;*
- *controlling the discharge of contaminants (pollutants) to air, land and all water including ground, fresh and coastal water; and*
- *managing the adverse effects (environmental externalities) of all activities using land, air or water.*

It established clear environmental responsibilities and requirements for assessment and planning. Every local authority and the national government is required to account for the effects of development, and must monitor the state of the environment with a view to adjusting activities accordingly (for example, ensuring that future applications for resource consents do not exceed environmental standards and objectives). As in the Netherlands, New Zealand has avoided a 'centralized' approach to planning for sectors (such as energy and transport) and such functions are devolved to local authorities. The Resource Management Act represents a very big 'stick' — indeed, it is probably the most comprehensive 'big stick' in the world — in that it sets integrated 'bottom lines' which operate across all environmental media. But the Ministry for the Environment does not have the resources to offer the number of direct financial 'carrots' that some other countries provide. But it does work with mechanisms such as voluntary agreements and uses its limited resources to maximize the 'carrot' dimension.

It needs to be acknowledged that, as with national strategies, progress with sub-national strategies can be elusive. Sadler (1996) describes the experience in British Columbia:

> *In 1992, the provincial round table released an assessment of the state of sustainability in British Columbia, concluding 'that, in may cases, present patterns of human activity and trends in expectations are not sustainable' (BCRTEE, 1992, p15). It also provided a comprehensive statement of the principles, criteria, tools and decision-making reforms necessary to turn this situation around. This was seen as a first step toward developing a strategy — a plan of action — that sets out clearly how sustainability can be achieved. Less than one year later, the BC round table was closed down, with unfinished business on the books.*
>
> *On the positive side, however, aspects of the sustainability agenda were incorporated into the mandate of the Commission on Resources and the Environment (CORE) ... [which is] developing a province-wide strategy for land use planning and resource and environmental management.*

Box 6.6 EXAMPLES OF SUB-NATIONAL STRATEGIES IN CANADA

The first provincial conservation strategy was developed in Prince Edward Island in 1987 by the Department of Environment, and Alberta and Quebec also began work in the mid-1980s. The Yukon Conservation Strategy, prepared by the Department of Renewable Resources in cooperation with a public working group, was released in 1990.

Provinces and territories in Canada all have developed, or are in the process of creating, sustainable development strategies. For the most part, these are the product of provincial round tables. They have been individually and comparatively reviewed in two recent reports (Weichel, 1993; Clement, 1993). Strategies have been completed in British Columbia, Saskatchewan, Manitoba, Ontario, New Brunswick and Nova Scotia. In Alberta, the round-table process led to a statement of sustainable development principles and the identification of some priority areas, as well as a vision statement. In Quebec, an eco-summit, based on regional activities, is being organized for the autumn of 1996. As another alternative, in the Northwest Territories, a sustainable development policy has been implemented.

There are also some important regional initiatives:

- *the Arctic Environment Strategy (a corollary of the Green Plan) focuses on cleaning-up the more than 800 hazardous solid waste dumps and industrial sites distributed across the region, emphasizes the provision of jobs and skills opportunities for northerners, and promotes community-based resource management. This aspect is extended in*
- *the Inuit Regional Conservation Strategy (1986), prepared by the Inuit Circumpolar Conference, in response to the World Conservation Strategy, and other Inuit-led initiatives;*
- *the work of the International Joint Commission (IJC), established under the Boundary Waters Treaty of 1909, particularly concerning the Great Lakes and St Lawrence river (water quality, and ecosystem restoration and protection);*
- *initiatives of the Atlantic Canada Opportunities Agency (ACOA), for example, cooperation agreements on sustainable economic development in Nova Scotia and Prince Edward Island; and*
- *various coastal and river action plans and management plans (for example, Fraser Basin Management Program, Gulf of Maine (marine) Action Plan).*

In the USA, some states have developed innovative environmental and resource programmes that are considered to offer models, pilot programmes, and demonstration projects for aspects of a national strategy (see Box 20.4). In developing the national sustainable development action strategy for the USA (PCSD, 1996), the President's Council on Sustainable Development held early discussion with those responsible for the Minnesota State strategy.

Local Agenda 21s

In operational terms, many of the objectives of sustainable development have to be negotiated locally. For it is here, in both rural and urban areas, that communities and individuals take daily decisions about resource use or environmental management — decisions that affect their livelihoods and often their survival. Redclift (1992) captures a growing consensus concerning the role of communities when he argues that:

> *sustainable development might be defined by people themselves, to represent an ongoing process of self-realization and empowerment' ... and that the 'bottom line', in practical terms, is that if people are not brought into focus through sustainable development, becoming both the architects and engineers of the concept, then it will never be achieved anyway, since they are unlikely to take responsibility for something they do not 'own' themselves.*

Dalal-Clayton et al (1994) have argued that it is vital that a national sustainable development strategy (NSDS) is not developed in isolation, at the national level only, and in a 'top-down' manner. They take the view that,

> *An NSDS process needs to be as fully participatory as possible, initially within political, social and cultural constraints that may exist. It will also need to be complemented by sub-national (regional/ provincial, district) and even more local-level strategies which will be best able to address the real and substantive issues which concern local communities.*

Over the last few years there has been an 'explosion' in all the industrial countries of Local Agenda 21s and similar initiatives. In some countries, the national processes have sought to promote, or have led subsequently to efforts to foster or support, local strategies. But in other countries, Local Agenda 21s have been initiated, developed and implemented independently of national strategies.

Local Agenda 21s Linked to National Strategies

In Australia, following the process of developing the *National Strategy for Ecologically Sustainable Development* (CoA, 1992a), the federal government established a Local Agenda 21 project, part of which included the preparation of information kits to assist local councils with implementing Agenda 21. The project, due to be completed in April 1994, provided a guide and included an education campaign (DEST, 1994).

Similarly, in France, following implementation of the *Plan national pour l'Environnement* (PNE) (French MoE, 1990a), the Ministry of Environment produced a guide to assist towns to develop environmental plans (identifying priorities) under environmental charters (*chartres pour l'environnement*) and also provides some funding and assistance. About 25 towns and several departments (provincial authorities) have signed contracts with the MoE to develop such plans. The aim is to undertake a 'global' (all embracing) environmental protection approach towards natural resources and fauna, landscapes, urban and rural heritage, waste elimination and the prevention of natural risks. Under the contract,

> *the state undertakes to contribute its technical and financial assistance to the local authorities who so desire. In turn, these agree to comply with the principles and goals*

of the Environmental Charter and to initiate a three-staged approach: environmental diagnosis, defining concerted priority goals, and drafting a programme of multi-annual actions.

(French MoE, 1995)

Following this pattern, the Ministry for the Environment in New Zealand has also developed a framework to guide local authorities to develop Local Agenda 21s and has provided some technical assistance. A number of cities have now developed local strategies (for example, Waitakere, Hamilton).

Comparable approaches have been adopted in the Scandinavian countries. Part 2 of Norway's *Report to the Storting No 46* on WCED follow-up (NorMoE, 1989) contains a chapter on how local processes should proceed. Each local community was asked to develop its own environment plan. The report announced that the government would provide finance for the appointment of an executive officer in every local authority, who would be responsible for environmental issues and would assist in the preparation of local environment plans. Several local communities are also preparing independent Agenda 21s and 'eco-community' plans.

In neighbouring Sweden, during preparations for Government Bill 1993/4.111, *Towards Sustainable Development in Sweden* (SwedMoE, 1994), all 285 local authorities were invited to embark on preparing Local Agenda 21s and 200 are now engaged in developing them. According to environment ministry staff, they represent 'more of a process than plans'. The separate *Enviro '93* strategy — Sweden's Environmental Protection Agency's action programme (SwedEPA, 1993) — is much used by communities preparing Local Agenda 21s and by county administrations developing their environmental policy plans, to set environmental targets.

During the preparation of Denmark's new *Nature and Environment Policy* (DanMoE, 1995) the Minister of Environment and Energy, together with the heads of the Association of Municipalities and the Association of Counties, wrote to the leaders of all counties and municipalities urging them to ensure that Local Agenda 21s were started. Some local municipalities/authorities are now developing such local strategies, but the Nature and Environment Policy Plan is not linked to these. However, it will be sent to all local authorities so that it can influence their Agenda 21 processes. The Ministry of Environment (MoE) is also encouraging them and is providing advice. A working group comprised of representatives of the Association of Municipalities has been established. The MoE, in collaboration with the Association of Municipalities and the Association of Counties, has produced a guide on preparing Local Agenda 21s (DanMoE, 1995a). This provides concrete examples and explains how to organize the process. It is targeted at local politicians and civil servants. The three organizations intend to organize seminars and conferences to discuss opportunities and to exchange experiences. Some of the MoE authors of the guide also contributed to writing the urban section and political introduction to the draft policy plan, providing some linkage between the two processes. Policy-making in Denmark is a very decentralized process, with considerable dialogue between civil servants and politicians at national, county and municipality levels.

In November 1994, a conference was organized in the Netherlands for representatives of the Ministry of Housing, Spatial Planning and the Environment (VROM) and of more local levels of government (such as provinces, municipalities and water boards) to discuss how the latter could implement the second *National Environmental Policy Plan* (NEPP2) (VROM, 1993a). Each province has its own environmental policy plan as a guide to carrying out its legal duties and obligations (for example, to check company permits), and municipalities have collectively developed an action plan to help them in their responsibilities to implement NEPP2. Water boards also have particular responsibilities to implement elements of NEPP2.

In Canada, following a *Green Plan* commitment to implement a marine environment programme (Government of Canada, 1990a), the Atlantic Coastal Action Plan was initiated to develop strategies or 'blueprints' for managing the coastal resources of 13 communities in Atlantic Canada.

Local Agenda 21s Initiated Independently of National Strategies

In Canada, there are also various 'independent' local level sustainable development initiatives (see Box 6.7). In Australia, Agenda 21 generally has a low profile at local government level, but some states and territories are promoting Local Agenda 21s. In South Australia, for example, a government officer has been appointed to assist local authorities in developing their own Local Agenda 21 plans, incorporated within their regular planning.

In the USA, a wide range of local initiatives has been launched. A well-known example is Sustainable Seattle — a volunteer network and civic forum committed to preserving the social, economic and environmental health of the Seattle area (Lawrence, 1994). In 1993, Sustainable Seattle published *1993 Indicators of Sustainable Community*, cited in Nguyen and Roberts (1994) and an indicators' task team was working on an updated version. Other projects include a community outreach project, a policy team to monitor government initiatives, and the marketing and communications team, which designs a communications strategy for the sustainability effort.

Another example is the Southern California Council on Environment and Development — a 'coalition of environmental organizations, citizens groups, government agencies and private enterprises dedicated to furthering sustainability'. The council works via a round-table process to implement recommendations of Agenda 21 in municipalities in southern California and to serve as a forum for promoting sustainable development throughout the region. Other projects include developing sustainability indicators and inventorying current policies, programmes and costs. A range of other local initiatives is described in Box 20.4.

Following UNCED, a number of cities and larger towns in the Netherlands have independently developed Local Agenda 21s (for example, Leiden) and these tend to be very participative. However, they are completely independent of NEPP2. In response to Agenda 21 and the EC's Fifth Environmental Action Programme, local government in the UK is actively developing its own policies and programmes (running parallel to the UK national strategy) through the 'Local Agenda 21 initiative' (see Chapter 19, section on provincial and local strategies, p191).

In 1993, under the Local Government Act, local authorities in the UK were given new legislative powers to participate in overseas assistance programmes. This has enabled local government to collaborate with southern partners in the promotion of policies and programmes that promote sustainable development practice (see Box 19.2).

Box 6.7 EXAMPLES OF 'INDEPENDENT' LOCAL STRATEGY INITIATIVES IN CANADA

Strategies have been prepared by several local or regional municipalities (for example, Vancouver City Plan — a participatory planning process launched in late 1992; Hamilton-Wentworth Vision 2000 — a sustainable development strategy for that region). Since 1987, local round tables have been appointed by various city councils. In Manitoba alone, there are 52 local round tables, approximately 40 per cent of which have completed sustainability vision statements or strategies (Sadler, 1996). Other local sustainable development initiatives and processes that complement the national ones include:

- *Remedial action plans (RAPs) being developed to restore and protect waterfront areas by 17 communities in the Great Lakes region (for example, Hamilton Harbour RAP).*
- *Eight (to date) self-help small town and rural community sustainability planning processes established since 1991 under Mount Allison University's Rural and Small Town Research and Studies Programme.*
- *Watershed-based management plans and projects (for example, 75 community-based watershed management projects on Prince Edward Island, and watershed advisory groups in Nova Scotia).*
- *Various programmes and projects aimed at 'greening communities' (focusing mainly on water and energy efficiency and conservation, water quality, waste reduction and management, greenspace planning, parks, natural areas, and wildlife habitat conservation).*

In eastern Europe, the position is variable. In Poland, there are several local strategy initiatives. For example, the Radom Project for Sustainable Communities involved training all key groups in risk assessment, and a series of participatory workshops which involved the setting down of priorities for environmental management within the municipality. But, from 1995, the preparation of Local Agenda 21s has become a legal requirement. Each *gmina* (local administrative body, typically comprising a dozen or so parishes) now has a legal obligation to develop a local sustainable development strategy (Local Agenda 21) within the next five years. In practice, there appears to be far more enthusiasm in Poland for local strategies than for national level ones. In Latvia, by comparison, no local strategies yet appear to have been initiated.

Convention Strategies

The conventions on climate change, biodiversity and desertification require signatory countries to prepare national plans outlining what measures are needed to deal with problems and meet obligations. It has been argued (Dalal-Clayton et al, 1994) that a national sustainable development strategy (NSDS) can form an umbrella of broad objectives, institutional roles, decision-making and monitoring processes, and guidelines under which more detailed strategies (local, sectoral and convention-related) should be formulated and implemented. However, in most countries, NSDSs or equivalent green plans have been developed independently of new convention strategies. Obviously, some strategies and green plans were developed before the conventions were negotiated, but other more recent ones still appear to be mainly de-linked from work on convention strategies.

Among the strategies studied, one exception appears to be Latvia, where the Nature Protection Division takes responsibility for developing the biodiversity action plan (BAP). Great effort is being made to ensure that the BAP is relevant to Latvia and that it ties in with the processes of the 1995 *National Environmental Policy Plan for Latvia* (NEPPL) and of the National Environmental Action Plan. The person responsible for the BAP is a member of the NEPPL core group and this allows for easy flow of information. In Latvia, as elsewhere in eastern Europe, ratification of conventions has serious financial, administrative and institutional implications and places a great burden on existing staff whose time is already limited.

Chapter 7 POLITICS, GOVERNANCE and LEGISLATION

Green planning and the development of sustainable development strategies have been influenced by a variety of factors, particularly commitments entered into at UNCED and under international agreements and treaties. But some strategies have also been strongly influenced by domestic political agendas.

Political Influences

In Canada, in the late 1980s, the Conservative government was faced with soaring public demands for action on the environment rather than simply rhetoric. There had been high-profile controversies with Alberta, British Columbia, Quebec and Saskatchewan over environmental assessments of major projects; several environmental disasters both in Canada and abroad; and considerable international pressure. The government's response was the *Green Plan* (Government of Canada, 1990a). Toner (1994) comments that:

> *the government's strategy was to launch a major well-funded initiative, implement it, gain political credit for responding to the public's desire for improved environmental quality and run on it in the next election. But it did not turn out that way. While the Tories found the political will to launch the Green Plan, they had difficulty sustaining their commitment because it was not based on a strong ideological or emotional foundation. By the middle of 1993, they had lost interest in the Green Plan as environmental issues declined in the 'top-of-mind' public opinion surveys.*

Political considerations also greatly affected the structure and content of the Green Plan. Early drafts were much closer to a sustainable development strategy than the final document. Toner (1994), in a critical review of the Green Plan experience, comments that:

> *The early drafts envisioned the Green Plan as representing a turning point in the Canadian discourse by moving the conceptual basis of environmental policy away from resource management and environmental clean-up to pollution prevention and sustainable development. In the end, the modifications required to get the final document through the Conservative cabinet diminished much of the Green Plan's vision and focus.*
>
> *Indeed, the drafts written in the autumn of 1989 identified the societal and economic decision-making systems as the 'root cause' of environmental degradation. This discussion of decision-making in the opening chapter provided the Green Plan with an analytical framework that structured the over 100 initiatives it introduced to address the various environmental problems. When the politicians on the Cabinet Committee on the Environment (CEE) undertook their detailed review of the draft Green Plan in the fall of 1990, they imposed a traditional 'distributive politics' template on the document by moving the expensive environmental clean-up programmes to the front and burying the chapter on the need to change societal decision-making systems in the back of the text. As a result, the final form of the Green Plan looked less like a novel, sustainable development strategy and more like just another environmental protection program. Even then, it eschewed greater reliance on the traditional regulatory approach. The ministers' decisions reflected both their own electoral calculus and the advice of their departmental officials. This change in the approach compromised the integrity of the Green Plan.*

The Conservatives were criticized for their 'eco-backtracking' and went on to lose the election in 1993. The incoming Liberals had been very critical of the Green Plan in opposition, arguing that it did not go far enough in institutionalizing a sustainable development framework. They dedicated a chapter of their election 'red book' (*Creating Opportunities: The Liberal Plan for Canada*, Canadian Liberal Party, 1993) to the theme of promoting sustainable development as an integral component of decision-making at all levels of society, and several features of their programme go beyond the promises made in the Green Plan. It is now pursuing this agenda. Green Plan programmes have been integrated into each department's activities and are no longer identified separately. Among other initiatives, legislation has been passed to establish a 'Commissioner of the Environment and Sustainable Development' (see Chapter 10).

The fortunes of Australia's *National Strategy for Ecologically Sustainable Development* (NSESD) (CoA, 1992a) have also been strongly influenced by political changes. As in Canada, public pressure and concern about the environment, and a need to resolve conflicts over resource use and environment–development issues — a problem faced by successive governments — triggered the development of NSESD. But support for the process from the Prime Minister, Bob Hawke, was an important element in ensuring its initial success. When Paul Keating assumed the premiership in 1991, he showed no interest in the NSESD and it lost momentum. Other factors also played a role, including:

- economic recession (with government refocusing on traditional job and growth issues);
- environment concerns assumed lower priority on public and political agendas; and

- the strategy process became a multifaceted look at sectors, and was therefore non-threatening to any particular sphere of government or business.

The origins of New Zealand's 1991 *Resource Management Act* (RMA) can also be traced to political changes. In 1984, the National Party government of Prime Minister Muldoon was replaced by a Labour administration. This was accompanied by a swell of public pressure in favour of environmental reform. The incoming government had made public commitments on the environment, which subsequently led to the establishment of the Ministry for the Environment (MfE). The new ministry staff recognized the need for fresh legislation to deal with the institutional problems and, in 1988, proposed the development of the RMA. This found resonance with the Environment Minister (Geoffrey Palmer) — a lawyer with a keen interest in law reform — and was endorsed by Cabinet with some modifications. Subsequently, a new National Party administration took office and, under Minister for the Environment Simon Upton's leadership, reviewed, refined and enacted what is now the RMA (Upton, 1995).

A change of government was also necessary in Sweden in order to establish a National Commission for Sustainable Development to take up sustainable development issues and to involve all actors and stakeholders. This was only achieved when the Social Democrats were elected to government in 1994. The previous government had not favoured such an approach.

In France, the *Plan national pour l'Environnement* (PNE) (French MoE, 1990a) benefited enormously from the personality and public standing of Environment Minister Brice Lalonde. Having started a Green Party in France in the 1970s, he had been one of the first people in Europe to introduce ecological issues into politics. The influence of the Green Party had increased over the years and M Lalonde had stood for President on several occasions. So the PNE was also a political response to growing public interest in the environment.

Work on preparing Denmark's new *Nature and Environment Policy Plan* (DanMoE, 1995) was triggered by a question in Parliament by the opposition Socialist Party on how the government intended to deal with strategic planning. In subsequently responding, the Minister of Environment formally announced an intention to develop strategic environmental planning, including policy and state of the environment reports for parliamentary discussion. The procedure was to build on several existing elements, which included regular ministerial sector strategies, as well as a range of periodic policy and state of the environment reports, and other policy-oriented papers. The parliamentary question enabled the MoE actively to promote consolidation and move further towards an integrated approach, with individual ministries retaining their responsibilities and being answerable to Parliament on how they were fulfilling environmental objectives in their areas of concern.

Political changes in central and eastern Europe, including the collapse of the former Soviet Union, have played a significant role in enabling progress with green planning and strategy development. For example, the development of Poland's *National Environmental Policy* (NEP) (PolMEP, 1990a) was the result of a series of events connected with the political situation in the country. One of the initiating forces behind the NEP process was growing awareness of the severe environmental damage suffered during the communist period and the need for radical policy reform. Repeated inaction and opposition by the communist regime led to the emergence of a strong environmental movement in Poland. The emergence of the Solidarity union in 1980 reinforced the environmental movement and provided a forum for public expressions of concern about the environment. The environmental movement grew in force and sought to inform the public of the linkages between environmental destruction in Poland, communism and central planning. This movement supports the notion that the government should integrate concerns about sustainable development into sectoral policies, and it managed to exert considerable influence during the 1989 round-table negotiations between the opposition and the communist regime, which led to the formation of a subgroup on the environment.

The unstable political situation in Poland has led to continual changes in cabinets, and has also led to reorganization within the Environment Ministry. This has prevented the development of stable policies and, with staff changes, has resulted in a loss of institutional memory. In 1990/1, the Environment Ministry was in a strong position, but is now one of the weaker ministries. It no longer has political weight and cannot, therefore, exert much political pressure on other sectors.

Cabinet and Parliamentary Review/Debate

Most strategies officially initiated by governments have been discussed and approved by cabinets or cabinet committees. Usually, but not always, they have been formally lodged as parliamentary documents and/or presented to parliaments or parliamentary committees. Most strategies tend to be policy documents with no legal standing as such (legal changes have usually been instituted through other instruments). Therefore, while a debate may have taken place on the strategy and it may have been endorsed, no formal vote has been necessarily required (for example, Canada's *Green Plan* (Government of Canada, 1990a)). However, where the strategy or green planning initiative has involved legislative changes (for example, New Zealand's *Resource Management Act* 1991), a full parliamentary debate and vote have been necessary. The position in respect of each strategy covered in this study is discussed in Chapters 9 to 21.

The position is complicated in federal states such as Australia and Canada where federal government strategies are not binding on individual provinces, states or territories. In Australia, the *National Strategy for Ecologically Sustainable Development* (NSESD) (CoA, 1992a) was endorsed by all heads of government at the Council of Australian Governments and subsequently approved by federal and state/territory cabinets. The Canadian Green Plan focuses on environmental action in areas under federal responsibility, and separate federal-provincial agreements were used to implement much of the plan (for example, in agriculture). Individual provinces are developing or have developed their own strategies.

Legislative and Institutional Consequences of Strategies

Only seven of the 20 strategies studied appear so far to have led directly to legislative changes (see Table 2.1).

The Australian *National Strategy for Ecologically Sustainable Development* (CoA, 1992a) included a proposal to create an Environmental Protection Council. An act giving effect to this was passed in 1994 and all state and territory governments have agreed to introduce mirror legislation. However, some observers have been very critical of the Australian federal government for not being prepared to take a strong leadership role in ensuring that the concept of sustainable development is woven into the fabric of federal environmental law (see Box 9.5), and for being content to:

> *leave legislating for sustainable development largely in the hands of the States and Territories. ... Each State and Territory has been proceeding down its own path in either implementing the concept or ignoring it. As a result, the ad hoc and uncoordinated development of environmental laws in Australia at Commonwealth, State and Territory level has persisted through the 1980s and 90s and looks set to continue.*
>
> (Scanlon, 1995)

In direct contrast, New Zealand's *Resource Management Act* (1991) was itself a major piece of reforming legislation which rationalized a confusing 'mess' of overlapping and contradictory resource legislation and responsibilities among various agencies. It set out to guide the sustainable use, development and protection of natural and physical resources. The subsequent *Environment 2010 Strategy* (NZMfE, 1995) includes various elements, including a new Fisheries Act, which embodies a sustainable development approach and a Hazardous Substances and Organisms Bill.

In Canada, in part as a follow-up to the *Green Plan* (Government of Canada, 1990a) and also as a response to UNCED, legislation has been passed by Parliament to establish a Commissioner of the Environment and Sustainable Development to hold the government accountable for 'greening' its policies, operations and programmes. The legislation also requires all federal ministers to table departmental sustainable development strategies in Parliament (see Chapter 10 for further details).

The French *Plan national pour l'Environnement* (PNE) (French MoE, 1990a) represents a set of proposals, two-thirds of which have been implemented by successive governments, in part through a range of new laws concerning such things as water and waste.

The Dutch *National Environmental Policy Plan* (NEPP) (VROM, 1989) was prepared as a result of a process later laid down in the 1993 Environmental Management Act, in which integrated legislation on the environment replaced previous sector-based laws. This required the NEPP to be revised every four years and extended to the regional and municipal levels.

In Norway, a range of ideas were incorporated in *Report to the Storting No 46 (1988–89), Environment and Development: Programme for Norway's Follow-Up of the Report of the World Commission on Environment and Development* (NorMoE, 1989). Some of these have taken several years to become adopted, for example, a change in the law to make pollution regulations apply to roads was due to be considered by the Cabinet in April 1995.

During the development of Sweden's Government Bill 1993/4:111, *Towards Sustainable Development in Sweden* (SwedMoE, 1994), many ideas on sustainable development were discussed and these were introduced into other bills in other sectors being developed at the same time. For example, the Ecocycle Bill (March/April 1994) — based on the concept of the 'ecocycle society' — adopted many arguments expressed in Agenda 21 and in Bill 1993/4:111. The latter, while not an act with the force of legislation itself, has influenced legislative changes and is frequently referenced. It led to the 1994 Finance Bill earmarking 10 million Kronor to support Local Agenda 21s (through experimental initiatives/projects). A strategy for biodiversity conservation was adopted by Parliament in 1993 (Bill 1993/4:30).

It is difficult to make clear judgements about institutional changes that have followed specifically as a result of strategy processes, rather than as a process of reforms that have proceeded for other reasons or under different stimuli. However, some direct institutional changes are clearly evident in particular cases. For example, the French *Plan national pour l'Environnement* (French MoE, 1990a) resulted in the reorganization and strengthening of the Ministry of Environment's structure. This was achieved through increasing the staff complement and budget, providing a more independent structure with regional offices and regional direction, and increasing the ministry's profile and influence. The PNE also led to the establishment of a French Institute for the Environment. In New Zealand, the *Resource Management Act* (1991) introduced a wide range of reforms, including, among others, the complete overhaul and rationalization of institutional responsibilities for environmental management (see Chapter 15).

In June 1993, the European Commission approved a set of internal measures to integrate the environment into all its policy and actions (see Chapter 21, pp230–2).

Chapter 8 CONCLUSIONS

Industrial countries have a long history of planning and, today, spatial planning systems tend to be institutionalized at all levels of government. Many of these systems are or could become important processes for promoting components of sustainable development (Sadler, 1996). Good examples include regional land use planning, resource management and environmental impact assessment, but they often fall short of realizing their potential because of the absence of an integrated policy context. They are also limited in their scope of application.

Towards Sustainable Development Strategies

Over the last few years, many industrial countries have undertaken or embarked on national planning exercises (green plans) to deal with growing environmental problems and increased public concerns about such issues. Following UNCED, most are now attempting to address the issue of sustainable development and are considering how to respond to Agenda 21 at a national level. National sustainable development strategies (NSDSs), called for in Agenda 21, have the potential to provide the framework for policy and institutional integration and, according to IUCN (1993c), to act as:

- a forum and context for dialogue on sustainability, including the guiding visions and values;
- a framework and process for identifying major issues and priorities;
- a mechanism for focusing policy and research to address these; and
- an approach to building capacity and strengthening institutional arrangements for implementing agreed upon actions.

Each country will need to determine, for itself, how best to approach the preparation and implementation of an NSDS. It is widely accepted that a 'blueprint' approach is neither possible nor desirable. Yet, despite all the rhetoric and agreements reached at UNCED, there is still considerable debate on just what sustainable development means in practice. No country has yet developed a green plan or strategy that analysts would agree meets all the requirements that might be expected of a genuine NSDS — even though some initiatives have been given such a label (for example, the *UK Strategy for Sustainable Development*, HMSO, 1994a). However, it would be unrealistic to expect this to be the case at present. Undertaking a strategic approach to sustainable development is arguably one of the most difficult challenges for any country to meet and all countries are still effectively 'learning by doing'. All of the initiatives reviewed in this study represent honest steps in this learning process undertaken by those involved. Many of the individuals responsible for coordinating and leading these initiatives remain open-minded about how far environmental planning can be pushed towards addressing sustainable development issues.

Among the strategies discussed in this book, with the exception of the Canadian *Projet de société* and *Action Plan: Sustainable Netherlands* (Milieudefensie, 1992) (both non-governmental initiatives), the processes followed have been fashioned mainly by prevailing political, bureaucratic and cultural circumstances in the industrial countries concerned, and have usually adopted approaches consistent with routine government practices for such initiatives. Furthermore, the different plans and strategies have been developed to address particular domestic environmental, social and economic conditions and circumstances, which differ in each country. While the initiatives covered can all be described — and indeed are promoted by their principal architects — as green planning processes, in practice they represent a range of quite different approaches (for example, environmental plans, strategies, legislative instruments, reports to Parliament, Commission processes). They are also aimed at fulfilling a variety of different objectives, some visioning, some goal-setting and some for implementation. They are not equivalent processes and it is impossible to compare them as if they were. This also makes it difficult to produce meaningful comparisons (beyond those in the preceding chapters) on how they have responded to the dilemmas listed in Chapter 1, or to judge whether they provide support for the key lessons and guiding principles for the development of genuine sustainable development strategies listed in Box 1.2.

Developed and Developing Country Approaches Briefly Compared

In approaching the challenge of creating national sustainable development strategies, it is fair to say that the countries of the North and South have much to learn from each other's experiences. Over the last 15 years, developing countries have had much experience in setting up national conservation strategies (NCSs), national environmental action plans (NEAPs), tropical forestry action plans (TFAPs) and many similar approaches. Several recent reports have reviewed this experience from a process point of view (for example, IUCN, 1993a and 1993b; Carew-Reid et al, 1994; Dalal-Clayton et al, 1994; ERM, 1994a, 1994b, 1994c, 1994d and 1995; Bass and Dalal-Clayton, 1995, OECD, 1995a; and World Bank, 1995).

This work has revealed that the more successful strategies in developing countries appear to share a number of commonalities (see Box 8.1). This is not surprising since many of the approaches have followed a basic framework developed for NCSs, which, as experience has grown, has been subsequently built on and improved for NEAPs, TFAPs and similar initiatives. Furthermore, these approaches have been mainly promoted by donors, who have provided the financial support and technical assistance to replicate the models in different countries as a framework for aid support. In many cases, the expatriate technical experts and advisers have worked on strategies in several countries and have transferred their experience and approaches.

But the situation in developed countries is entirely different. No common approach is apparent in the processes adopted — there are no common actors and there is no

Box 8.1 ***KEY TASKS IN THE STRATEGY PROCESS IN DEVELOPING COUNTRIES***

- Determine if conditions are appropriate—*for example, a conducive political and social climate, high-level political support, and adequate funds.*
- Decide on an entry point. *An NSDS should be a cyclical process. Some elements follow one from the other; others (for example, information analysis, monitoring and evaluation, and some implementation) proceed throughout the cycle. A new strategy should take account of what has gone on before, perhaps starting at whatever stage a significant ongoing or past strategy has reached.*
- Establish an engine to drive the process. *Often, a secretariat is formed, comprising committed staff with good management skills, both from inside and outside the government. The secretariat may be responsible to a steering committee with broad representation, and frequently independently chaired. Neither body should have vested sectoral interests or be located in a sector or interest group.*
- Decide the process design. *The steering committee and secretariat will need to determine the scope of the strategy, the main 'stakeholders' to be involved, the issues to address, the approach, and how to manage the individual elements that comprise the strategy cycle.*
- Determine the participants. *Participation implies full involvement of relevant groups (both government and non-government) in appropriate tasks including strategy design, exchanging information, decision-making and implementation. It is necessary to decide how much participation is possible and necessary, and to develop mechanisms for participation, for example, core groups, round tables, workshops and community-based techniques.*
- Information assembly and analysis. *This can be undertaken through background studies and workshops, and by government agencies, universities, research and policy institutions and independent professionals.*
- Policy formulation and priority-setting. *Establish principles, goals and objectives of the strategy, and targets for achieving objectives, through appropriate fora, for example, policy dialogues and round tables.*
- Address the hard questions of sustainable development. *The major issues, obstacles and risks will be subject to differing opinions and attitudes. There are likely to be winners and losers and trade-offs will be necessary. Policy dialogues should first focus on potential win-win situations, later moving to the more intractable issues.*
- Action planning and budgeting. *An NSDS is a 'macro' approach that needs on-the-ground 'micro' actions. These can include policy, legislative, institutional and organizational changes; capacity-building for government, NGOs and local communities; and a range of programmes and projects.*
- Implementation and capacity-building—*embracing the corporate sector, NGOs and communities, as well as government. Government can create an 'enabling environment' for development action by all sections of society, and NGOs can play a key role in catalysing participation and local action.*

- Communications — *keeping participants informed of progress, expressing consensus, generating wider understanding of sustainable development, and encouraging participation — through briefings, newsletters and media coverage.*
- Monitoring and evaluation — *of both the process and products.*

Source: Dalal-Clayton et al, 1994.

international protocol or format. As noted above, they have all been fashioned according to domestic agendas and have followed national government styles and cultures rather than those of external agencies. It is still too early to say whether any of the basic requirements that appear to characterize strategies in developing countries apply to those in developed countries. For example, it is logical that green plans and strategies in industrial countries should move closer towards the 'ideal' of sustainable development strategies if they are cyclical, i e if they are periodically revised to take into account feedback and lessons from review following implementation, and thus become genuine 'learning by doing' processes.[1] But to date, of the initiatives reviewed in this study, the only genuine second-generation processes are the second Dutch national environmental policy plan, NEPP2 (1993) — which built on the 1989 NEPP and NEPP3, planned for 1997, will presumably build further on this experience — and the European Union's Fifth Environmental Action Programme, which also built on previous programmes.

Despite these differences, some basic comparisons between the approaches followed in developed and developing countries can be made (see Table 8.1).

The aims of green plans and strategies in developed and developing countries appear to have been influenced by the usually quite different problems they have faced. Most developing countries are occupied with achieving economic development through industrialization, where possible, and by expanding production. By comparison, one of the key issues for sustainable development in most developed countries is dealing with the problems caused by high levels of consumption, by existing industries and by technology-based economies (such as pollution and waste). The transitional economies of central and eastern Europe (CEE countries) face urgent and complex environmental problems following years of ineffective central planning, poorly controlled industrialization and widespread ecological deterioration.

In developing countries, strategies have tended to lead to the creation of new institutions (such as environmental ministries or departments) and the introduction of legislation and procedures (for example, environmental impact assessment). However, as Sadler (1996) notes, industrial country responses to sustainability strategies have involved: (i) amendments to existing policy, legislation and institutional frameworks; (ii) a (sometimes linked) series of new initiatives; and (iii) a comprehensive strategy or plan implemented through various elements of (i) and (ii). A further comparison is that while industrial countries have used green plans and strategies to generate cost-saving approaches to environmental management, strategies in developing and CEE countries

1. This would match the experience of the private sector in industrial countries, which *has* taken a cyclical, continuous-improvement approach through adopting the processes set out by the International Standards Organization (ISO) and environmental management system processes.

Table 8.1 Comparisons Between Developed and Developing Country Strategy Processes

Developed Countries	Developing Countries
Approach	*Approach*
Internally-generated	External impetus (IUCN, World Bank, etc)
Internally-funded	Donor-funded
Indigenous expertise	Expatriate expertise frequently involved
Political action	Bureaucratic/technocratic action
Brokerage approach	Project approach
Aims	*Aims*
Changing production/consumption patterns	Increase production/consumption
Response to "brown' issues (eg pollution)	Response to 'green' issues/rural development
Environment focus	Development focus
Means	*Means*
Institutional reorientation/integration	Creation of new institutions
Production of guidelines and local targets	Development of project 'shopping lists'
Cost-saving approaches	Aid-generating approaches
Links to Local Agenda 21 initiatives	Few local links
Awareness-raising	Awareness-raising

have frequently been used as a mechanism to lever up additional aid support, usually through 'shopping lists' of projects.

Local Agenda 21 initiatives developed by municipalities and local communities are now common in most industrial countries and, in practice, are leading to significant levels of concrete local action and support. In many cases, such local initiatives are supported, or at least encouraged, by the national-level green plans and strategies. In developing countries, however, Local Agenda 21s are still relatively uncommon and local-national links tend to be few or weak. In developed and developing countries alike, awareness raising and education about environmental issues and sustainable development concerns have usually featured as primary concerns.

Challenges

In the preceding chapters, approaches to developing green plans and strategies for sustainable development in some selected industrial countries have been discussed. The analysis draws from the experiences over the last few years of 20 initiatives in ten countries, based on structured interviews with the key individuals responsible for managing those processes. The study complements other recent reviews by IIED and IUCN, which have focused mainly on the strategy experiences of developing countries (for example, Carew-Reid et al, 1994; Dalal-Clayton et al, 1994; IUCN, 1993a and 1993b).

While the green plans and strategies of many industrial countries are predominantly environmental in focus, undertaken to deal with pressing environmental problems and public concerns, others have been a direct response to agreements reached at UNCED, notably Agenda 21. From the experiences of both industrial and developing countries, however, it is evident that a holistic approach needs to be adopted in devising a strategy that genuinely addresses issues of sustainable development and that, where possible, it should seek to integrate environmental, social and economic objectives.

The mainly 'environmentally-focused' green plans covered in this book, many of which were initiated before UNCED in 1992, have laid foundations on which subsequent sustainable development strategies can build. But almost all of them, and many of the new breed of post-UNCED plans and strategies, have been developed through traditional policy-making practices, which are essentially internal to government and cannot deal effectively with the complexities of sustainable development. The goals of sustainable development should not be those of government alone. A bolder response is required, with more effort devoted to establishing interactive, participatory, strategic approaches.

The issue of participation is perhaps one of the most significant challenges faced by those responsible for strategies in developed and developing countries alike. For some countries, particularly those of central and eastern Europe, the idea of participation is still alien to governments and publics alike. As Bass et al (1995) have argued:

> *taking participation into the mainstream of planning and development activity needs further research and interaction, and changes in institutional and professional attitudes and environments. For these and other reasons, networking within and between countries on participation aspects is strongly recommended. This would be particularly valuable among policy analysts, planners and others who are working in multidisciplinary ways, and those who have been behind effective participatory approaches.*

Fora such as the International Network of Green Planners can play a valuable role in this regard.

While it is easy to call for a more holistic and integrated approach, we need to remember that, throughout the world, policies have rarely been integrated deliberately. Such deliberate strategic planning has always been difficult. Those policies that have been successful have tended to evolve over time. Sometimes policies and institutional arrangements and responsibilities overlap or are in conflict. In such cases, a strategy process can help to resolve matters. New Zealand's *Resource Management Act* (1991) was developed precisely to address such a problem.

A serious question, which will need to be faced by industrial countries if they are to make progress towards addressing the challenges of sustainable development, is to what extent will it be necessary to adopt the approaches found to be successful in developing countries (i e as suggested by the key tasks listed in Box 8.1)? In particular, to what extent will it be necessary to move towards more participative, integrative and cyclical processes?

Another question is whether governments are serious about moving their societies and economies towards a sustainable future, or are they merely paying 'lip service' and responding to the issue in a traditional way by driving Agenda 21 into an 'environmental rut'? While some of the initiatives discussed above have amounted to little more than environmental planning and policy-making as usual, others have made impressive progress towards building a sustainable future.

PART II
COUNTRY CASE STUDIES

Chapter 9 AUSTRALIA

Barry Dalal-Clayton and Barry Sadler

In Australia, the main national green planning process is the *National Strategy for Ecologically Sustainable Development* (NSESD) implemented by the Commonwealth government's Environment Strategies Directorate. The NSESD was initiated in 1990 by the Commonwealth Cabinet, and the strategy was drafted and endorsed in 1992 (CoA, 1992a); it is an ongoing process. An intergovernmental committee on ESD, comprising Commonwealth, state and local agencies, meets approximately quarterly to work through key sustainability issues, and to try to promote a national approach — which is critical in a federal system where responsibility for environmental and economic management is divided. The NSESD has a long-term vision, although not over any prescribed period of years.

The NSESD is the 'principal document' (*primus inter pares*) in Australian sustainability, guiding those aspects of government policy relevant to sustainable development. But there are various other important strategy or equivalent processes, the main elements of which are outlined in Box 9.1. This range of strategies may be seen in negative terms as a number of overlapping and potentially conflicting initiatives, or positively as a family of 'targeted' strategies.

The number of strategies also reflects the geopolitics of federalism in a country where individual states or territories have significant resource management responsibilities (as in Canada). The NSESD led to the Intergovernmental Agreement on the Environment (IGAE) (1992) — a political accord between the Commonwealth, each state/territory, and the Australian Local Government Association (ALGA) — an umbrella group for the 918 local councils (see Box 9.1). The IGAE is currently undergoing operational review.

In effect, there is a 'hierarchy of strategies' that are either serially or vertically linked (in practice, the degree of interrelation is unclear). For example, the national greenhouse response strategy (CoA, 1992b) was drafted parallel to the NSESD and these processes are best seen as a pair. Other strategies have also evolved on such a dual basis, for example, the National Forest Policy Statement.

Box 9.1 ***MAIN ELEMENTS OF STRATEGY PROCESSES IN AUSTRALIA***

Intergovernmental Agreement on the Environment (1992)

- *covers all levels of government, concerning roles and responsibilities for environmental decision-making;*
- *sets out principles and mechanisms for resolving national issues;*
- *based on acceptance of ESD principles;*
- *includes nine schedules dealing with issues such as land-use planning and approvals, EIA, climate change, and heritage conservation; and*
- *establishes a National Environmental Protection Council.*

National Greenhouse Response Strategy (CoA, 1992b)

- *endorsed by all Australian governments in 1992;*
- *outlines a range of measures for progress toward stabilizing greenhouse gas emissions at 1988 levels by 2000 and reducing emissions by 20 per cent by 2005; and*
- *covers both 'source' and 'sink' components, ie minimizing emissions and enhancing environmental capacity to absorb emissions (for example, carbon sequestering).*

Draft National Strategy for Conservation of Australia's Biological Diversity (ANZEEC, nd)

- *Australia ratified the UN Convention on Biological Diversity in June 1993;*
- *draft strategy sets out national goals and principles for meeting this commitment;*
- *recognizes the national forest policy strategy as a primary means of conserving biodiversity in forest habitats; and*
- *references other strategies.*

National Waste Minimization and Recycling Strategy (in preparation)

- *addresses the management of toxic chemicals and hazardous wastes;*
- *contains a series of reduction and recycling targets, for example, 50 per cent reduction in rubbish to landfill by 2000;*
- *incorporates a waste management hierarchy of (in order of preference) avoidance, reduction, recycling or reclamation, treatment and disposal; and*
- *complements a range of other national strategies and programmes including:*
  - *the draft national hazardous waste management guidelines;*
  - *a national strategy for management of chemicals used at work; and*
  - *a planned national pollution inventory.*

National Forest Policy Statement (CoA, 1992c)

- *signed by federal and state/territory governments (except Tasmania) in December 1992;*
- *sets out the government's approach to ESD for forests, including its relationship to national strategy and other policy initiatives;*
- *recognizes the unique character of Australian forests;*
- *contains a mix of policy positions and specific actions to be undertaken by signatories; and*

- *provides for a regional Commonwealth–state agreement for forest management, based on joint comprehensive assessment of forest values.*

National Water Quality Management Strategy (ARMCANZ/ANZECC, in preparation)

- *gives national guidelines for water quality management including:*
  - *sewage systems;*
  - *reclaimed water; and*
  - *urban storm water;*
- *emphasizes a package of complementary measures, including regulation, economic incentives and instruments and education;*
- *includes Australian guidelines for drinking water quality; and*
- *contains a monitoring and reporting component (linked to the National Waterwatch Programme which,* inter alia, *encourages community involvement and action in catchment management and landcare).*

National Landcare Programme (Campbell, 1992, 1994)

- *land degradation, including desertification, is of critical concern in Australia because of the arid nature of much of the country;*
- *wide range of policies and programmes aimed at promoting sustainable agriculture and repairing damage from past practices and land clearing activities;*
- *Landcare programme seeks to move towards an integrated, systemic approach rather than addressing individual resources and specific issues;*
- *it is described as a community-based, self-help initiative that focuses on socio-economic causes rather than the resource symptoms of degradation; and*
- *brings together a number of existing initiatives (for example, the National Soil Conservation Programme, and Save the Bush — i e remnant native vegetation, and the One Billion Trees Programme — replanting and restocking).*

The National Strategy for Ecologically Sustainable Development (NSESD)

Prime Motivation and Getting Going

Over the last 10 to 15 years, governments in Australia have taken more interest in the environment. The NSESD evolved over several years and through extensive consultation with all levels of government, business, industry, academia, voluntary conservation organizations, community-based groups and individuals. Its origins stem from the release of the World Conservation Strategy (IUCN/UNEP/WWF, 1980), the National Conservation Strategy for Australia (ANZEEC, undated), and subsequently the report of the World Commission on Environment and Development — the Brundtland Commission (WCED, 1987). The principles of ESD were first elaborated in Australia by the Australian International Development Assistance Bureau in the publication, *Ecologically Sustainable Development in International Development Cooperation: An Interim Policy Statement* (AIDAB, 1991).

The initiation of the NSESD can also be traced to the cooperative approach between the Farmers' Federation and the Australian Conservation Foundation to tackle land degradation — Landcare. These organizations proposed a strategy approach in outline to the federal government. In June 1990, the Commonwealth government set about the task of identifying comprehensively and systematically what Australians needed to do to embrace ecologically sustainable development (ESD), by instituting a process of detailed discussion involving governments and the community following release of *Ecologically Sustainable Development: A Commonwealth Discussion Paper* (COA, 1990).

The domestic trigger for the NSESD process was the need to respond to public pressure and to find a way through the resource use and environment — development conflicts encountered by successive governments (for example the Franklin dam in Tasmania; the wet tropics World Heritage site dispute in Queensland; mining in Kakadu National Park). Internationally, impetus was subsequently provided by the UNCED process. A key to the initiation and early success of the NSESD process was prime ministerial involvement (which initially took environmentalists by surprise). Some observers identify the NSESD closely with the consensus approach ('cooperative federalism') of Prime Minister Hawke — a former trade unionist. It was also a response to a groundswell of public opinion for action to deal with environmental issues.

The NSESD process was clearly driven by a unique political conjunction, in which the prime minister (Bob Hawke), a powerful environment minister (Senator Richardson) and certain 'green-minded' industry ministers made common cause. However, the underlying reasons for this are subject to varying interpretations — for example, the NSESD was 'an initiative designed by the prime minister to keep a powerful environment minister in check'. Whatever the machinations or motivations, the initial NSESD process represented an apparent change in political culture and the apparent emergence of a broadly-based constituency of support for its continuation.

Focus

> *The NSESD sets out the broad strategic and policy framework under which governments [states and territories] will cooperatively make decisions and take actions to pursue ESD in Australia. It will be used by governments to guide policy and decision-making, particularly in those key industry sectors which rely on the utilization of natural resources.*
>
> (CoA, 1992a)

ESD objectives and principles are clearly set out in the strategy document (Box 9.2). These are noble goals, with a strong sustainability-orientation. The NSESD is about broad policy integration rather than day-to-day, issue-to-issue, trade-offs (Box 9.3).

A key to 'unpacking' the NSESD as a policy concept is the participation of development interests in the process. In retrospect, development interests appear to have 'used' the NSESD process both to respond to environmentalist pressures in a concessional way and to try and appropriate the process so as to minimize its wider (economic) impacts. An important implication of their involvement was to reinforce the view of ESD as the green or environmental component of government policy. This is also reflected in and reinforced by parallel or subsequent sectoral initiatives, for example, in agriculture (the Landcare programme) and mining.

Box 9.2 GOAL, CORE OBJECTIVES AND GUIDING PRINCIPLES OF THE AUSTRALIAN NATIONAL STRATEGY FOR ECOLOGICALLY SUSTAINABLE DEVELOPMENT (NSESD)

Goal

- *Development that improves the total quality of life, both now and in the future, in a way that maintains the ecological processes on which life depends.*

Core Objectives

- *to enhance individual and community well-being by following a path of economic development that safeguards the welfare of future generations;*
- *to provide for equity within and between generations; and*
- *to protect biological diversity and maintain essential ecological processes and life-support systems.*

Guiding Principles

- *decision-making processes should effectively integrate both long- and short-term economic, environmental, social and equity considerations;*
- *where there are threats of serious or irreversible environmental damage, lack of full scientific certainty should not be used as a reason for postponing measures to prevent environmental degradation;*
- *the global dimension of environmental impacts of actions and policies should be recognized and considered;*
- *the need to develop a strong, growing and diversified economy which can enhance the capacity for environmental protection should be recognized;*
- *the need to maintain and enhance international competitiveness in an environmentally sound manner should be recognized;*
- *cost-effective and flexible policy instruments should be adopted, such as improved valuation, pricing and incentive mechanisms; and*
- *decisions and actions should provide for broad community involvement on issues which affect them.*

These guiding principles and objectives need to be considered as a package; one should not predominate over the others. A balanced approach that takes them all into account is required to pursue the goal of ESD.

Source: CoA (1992a)

Organization and Management

The NSESD document (CoA, 1992a) is one of the few NSDS documents that clearly describe the process of developing the strategy (Box 9.4). Responsibility for implementing the NSESD rests primarily with individual jurisdictions and agencies, with the Commonwealth Department of Environment, Sport and Territories (DEST) taking on

Box 9.3 *AUSTRALIAN NATIONAL STRATEGY FOR ECOLOGICALLY SUSTAINABLE DEVELOPMENT (NSESD) — MAIN CONTENTS*

Part 1: Introduction *Explains ecologically sustainable development, sets out Australia's goal, core objectives and guiding principles for the strategy (see Box 9.2), explains who will be affected, describes how the strategy has been developed, indicates linkages between the strategy and other government policies and initiatives, and introduces an accompanying 'Compendium of ESD Recommendations'.*

Part 2: Sectoral Issues *Provides the broad strategic framework for those key industry sectors which rely on natural resources as their productive base. A Challenge Statement, Strategic Approach and Objectives are given for eight sectors: agriculture, fisheries, forests, manufacturing, mining, urban and transportation, tourism, and energy.*

Part 3: Intersectoral Issues *Provides a Challenge Statement, Strategic Approach and Objectives for 22 crosscutting themes which are relevant to actions in several of the key industry sectors, including:*

- *Biological Diversity*
- *Nature Conservation System*
- *Native Vegetation*
- *Environmental Protection*
- *Land Use Planning and Decision-Making*
- *Natural Resource and Environment Information*
- *Environmental Impact Assessment*
- *Changes to Government Institutions and Machinery*
- *Coastal Zone Management*
- *Water Resource Management*
- *Waste Minimization and Management*
- *Pricing and Taxation*
- *Industry, Trade and Environment Policy*
- *Aboriginal and Torres Strait Islander Peoples*
- *Gender Issues*
- *Public Health*
- *Occupational Health and Safety*
- *Education and Training*
- *Employment and Adjustment*
- *Australia's International Cooperation and Overseas Development Assistance Policy*
- *Population Issues*
- *Research, Development and Demonstration*

Source: CoA (1992a)

Box 9.4 THE NSESD PROCESS

In August 1990, the former Prime Minister, Bob Hawke, announced the establishment of nine ESD working groups with wide representation including government officials, industry, environment, union, welfare and consumer groups, to examine sustainability issues in key industry sectors. Their purpose was to provide future advice on them.

Community consultation formed an important part of this process, with a series of one-day consultation forums being held around Australia to discuss mechanisms for integrating economic and environmental concerns, and opportunities for broader community comment on the interim reports of the working groups.

In November 1991, the nine ESD working groups produced reports covering agriculture, forest use, fisheries, manufacturing, mining, energy use, energy production, tourism and transport. In January 1992, the three chairs of the working groups presented further reports on intersectoral issues and 'greenhouse issues'. In all, 11 reports contained over 500 recommendations on ways of working towards ESD.

The ESD working group process was valuable in two key respects. First, it produced wide-ranging and innovative recommendations for action both within and across key sectors of activity. While unanimity was not reached in a number of areas, many of the recommendations had a wide measure of support from all the interests represented. Second, and equally important, it promoted a continuing dialogue between interests and community groups. As a result, there is a better understanding of the factual basis of the debate and a greater willingness from the broad range of participants to encourage action which takes account of all the interests involved.

The reports of the ESD working groups provided the foundation on which state and federal governments developed the strategy. In November 1991, heads of government agreed on a cooperative intergovernmental process for examining the recommendations of the ESD reports. They established the intergovernmental ESD Steering Committee (ESDSC) to coordinate the assessment of the many recommendations and their implications for current and future government policies, and to report to heads of government on the outcomes of these considerations.

In May 1992, heads of government also agreed to release a draft of the strategy as an officials' discussion paper, to promote discussion and obtain community views on possible future policy directions. This was primarily in recognition of the nature, range and significance of many of the issues covered by the ESD Working Group Reports' recommendations. The draft strategy was subsequently released by the Prime Minister on 30 June 1992 for a two-month public comment period.

Over 200 submissions were received in that period. The majority of these submissions advocated acceptance of, and mechanisms for implementation of, the final recommendations from the ESD working groups and chairs; the clearer identification of priorities and of agencies responsible for implementation; and clarification of the linkages between the Strategy and other government policies and initiatives. The change in structure and content in finalizing the Strategy were largely in response to these comments. The ESDSC also found the submissions a valuable source of information on the broader community's priorities, and utilized them in helping to determine final policy positions. In addition, the ESDSC produced a Compendium of ESD

Recommendations as an accompanying document to the Strategy. The Compendium describes, in tabular form, how the Strategy and the National Greenhouse Response Strategy, agreed by governments, together with examples of relevant existing policies, relate to the over 500 ESD recommendations.

At its meeting on 7 December 1992, the Council of Australian Governments endorsed the National Strategy for Ecologically Sustainable Development (NSESD), noting that implementation would be subject to budgetary priorities and constraints in individual jurisdictions.

The Council agreed that the future development of all relevant policies and programmes, particularly those which are national in character, should take place within the framework of the NSESD and the Intergovernmental Agreement on the Environment (IGAE) which came into effect on 1st May 1992. The Council encouraged business, unions and community groups to use the NSESD as a basis for actions which contribute to the pursuit of Australia's national goal for ESD.

(CoA, 1992a)

The NSESD states that the ESD steering committee, with representatives from all levels of government, will monitor the performance and development of the Strategy by coordinating a report to heads of government one year after its endorsement on initial progress in implementation, with a further report every two years thereafter, or as called upon by heads of government, on matters which need to be drawn to their attention. A first report has been produced, based on information provided by the states and territories (excluding Western Australia, which did not participate on the ESDSC during the reporting stage), relevant Commonwealth departments and agencies, the Australian Local Government Association (ALGA) and relevant ministerial councils. Each of these bodies prepared a report on the implementation of the NSESD in their area of responsibility.

(CoA, 1992; ESDSC, 1993)

an overall monitoring and coordination role. Key tasks are to coordinate responses internally within the environment portfolio; and to coordinate positions across the Commonwealth government through a committee structure.

In the initiation phase, the NSESD process engaged senior people, but this is no longer the case. For example, in South Australia, an adviser to the state cabinet was given responsibility for NSESD issues. But this position has now been abolished and NSESD responsibility now lies with a junior officer in the state's Department of Environment and Natural Resources.

In addition to DEST, two other agencies are currently active in the NSESD process: the Department of Primary Industries and Energy, and the Department of the Prime Minister and Cabinet. Central and line agencies all have environment units, which, rather than working to ensure that policies and decisions are consistent with the NSESD framework, appear to serve liaison and watchdog functions. A sceptic might see these units primarily as a means to 'keep the lid on the green debate, preventing it from overflowing into economic terrain'.

Terms of Reference

The terms of reference for the ESD working groups (which initiated the strategy process) were set out in 'charter letters' (dated 29 August 1990) from the Prime Minister (Bob Hawke) to the chairs of the three subcommittees into which these groups were organized (CoA, 1991, pp259–61). The letter referred the ESD chairs to two 'baseline' documents: (i) The *ESD Discussion Paper* released by the Commonwealth government; and (ii) a related treasury paper, *Economic and Regulatory Matters for Ecologically Sustainable Development*. It also set out fundamental goals and principles — to which the then government was firmly committed — to guide the working groups (see Box 9.2). The chairs of the working groups were responsible to the Cabinet's sustainable development subcommittee, chaired by the Prime Minister.

Participation

The introduction or development phase of the NSESD was very successful, characterized by a highly visible, intensive and interactive process in which a wide range of sectors and groups with traditionally competing interests worked together. Participation was a key element in the design of the strategy (i e the working group and public consultation process — see Box 9.4). However, public involvement has not continued. Because the ESD process has now been internalized within government, there are no evident mechanisms for further public dialogue. By contrast, other strategies have negotiating mechanisms built in (for example, the round table on climate change).

NGO, Public and Political Reactions

At the outset, NGOs and the public were enthusiastic and were provided with resources with which to participate. By the end of the process, many were disappointed and disillusioned by what was seen as having been a 'talk fest' with an 'inoffensive output' containing little of consequence. It had started with the best of intentions but had become 'watered down' and lost its momentum. The initial high hopes had now turned to cynicism about the federal government's commitment.

Some observers are critical of the Commonwealth government for having left sustainable development legislation in the hands of states and territories; they argue that it should now take a stronger leadership role and should set about overhauling the country's legislative framework in line with the concept of sustainable development (Box 9.5). Critics consider that the balance of policy development and action through the NSESD, the Intergovernmental Agreement on the Environment (IGAE) and the National Environmental Protection Council Act are still inadequate (see also sections below on key issues and consensus).

Key Assistance Factors

A number of important factors have assisted the NSESD process: prime ministerial and cross-government political support; public support for environmental reform; international developments, i e the UNCED process; resources made available to NGOs (the major associations) to participate; and the willingness of the business sector to participate.

Box 9.5 A VIEW ON THE NEED TO LEGISLATE FOR SUSTAINABLE DEVELOPMENT IN AUSTRALIA

Environmental lawyer, John Scanlon, writes:

Achieving sustainable development will require a change in attitudes and a reorientation of policies and institutions. Four principal instruments will need to be used to ensure that economic activity is constrained within ecological limits, namely, voluntary mechanisms, regulation, government expenditure and financial incentives, and it will be necessary to utilize each one of them.

Environmental laws that establish a clear framework within which policies can be developed to implement sustainable development, capable of determining the ecological constraints to economic activity, are essential. New Zealand has been a world leader in undertaking major structural reform of its environmental laws in an attempt to legislate for sustainable development. Environmental, economic and social differences will mean that no one legislative scheme will have universal application. However, the essential framework of the New Zealand Resource Management Act (1991) is capable of universal application.

While it has the constitutional power to do so, the Australian Commonwealth Government has not been prepared to take a strong leadership role in ensuring that the concept of sustainable development is central to the development of environmental law in Australia. Despite all the rhetoric, it has been content to leave legislating for sustainable development largely in the hands of the states and territories. ... Each State and Territory has been proceeding down its own path in either implementing the concept or ignoring it. As a result, the ad hoc and uncoordinated development of environmental laws in Australia at Commonwealth, State and Territory level has persisted through the 1980s and 90s and looks set to continue.

The current state of environmental laws in Australia is unsatisfactory and radical change is needed. That will require the Australian Commonwealth Government to take a strong leadership role in developing a legislative framework for Australia based upon the concept of sustainable development. The time has clearly arrived for the Commonwealth Government to stop just applauding this clever scheme called sustainable development and to take the brave step of putting the bell on the cat.

Scanlon also argues that the Australian Commonwealth government 'has not genuinely sought to weave any of the following fundamental issues into the fabric of environmental law within Australia:

- *recognition of the concept of sustainable development;*
- *determining ecological limits to economic activity;*
- *determining the role and function of different spheres of government;*
- *integrating relevant environmental laws;*
- *providing for effective public participation;*
- *making decision-makers accountable for their actions; or*
- *recognizing and strengthening the role of indigenous people and their communities.*

Source: Scanlon, 1995.

Other mechanisms have helped to support the NSESD, but largely in a peripheral manner. For example, environmental impact assessment serves as a 'safety net' to catch major developments downstream. There appear to be few proactive instruments for capturing these issues at the policy level, for example a policy review and audit capacity are critically needed if the causes of unsustainable development are to be addressed (see section below on conflict resolution).

Key Problems

The early promise of the NSESD (see section on prime motivation and getting going, p77 above) has not been sustained, and the momentum of and government commitment to the NSESD process has been lost before real change has occurred. Some of the apparent reasons for this include:

- changing political leadership (Bob Hawke promoted consensus-building 'cooperative federalism' and was a champion of the NSESD process; Paul Keating, who succeeded him as Prime Minister in December 1991, followed a different and more strident philosophy of 'new federalism');
- economic recession (with government refocusing on traditional job and growth issues);
- environment concerns have assumed a lower profile than they had at the start of the process; and
- the strategy process became a multifaceted look at sectors and was therefore non-threatening to any particular sphere of government or business.

In addition, sustainable development is hard to maintain as a concrete policy focus, compared with specific 'green campaigns' such as saving a forest or recycling wastes. It encompasses a range of cumulative issues, which are widespread and pervasive but also incremental and insidious, so often overlooked. In Australia, drought, soil loss and salinization are major long-term, broad-scale issues, which collectively impose a serious constraint on resource productivity and agricultural sustainability (see National Landcare Programme in Box 9.1).

The sheer number of issues covered in the NSESD was symptomatic of the difficulties encountered in policy response. Approximately 500 were identified and, although these were grouped into sectoral and crosscutting areas, there was no actual focus on what options could make a real difference (for example, influence on the greening of the economy — see section below on greening the mainstream). Ironically, by dealing with all the issues (as advocates of sustainable development argue is necessary), sustainability becomes difficult to 'policy manage', and taxes conventional institutional arrangements.

Key Issues

The ESD working groups addressed a variety of key issues: agriculture, forest use, fisheries, manufacturing, mining, energy use, energy production, tourism and transport. The key one now is whether and how the NSESD can be renewed and/or restructured. It is significant that the current framework fails to provide leverage on the major economic issues and that, from a political and bureaucratic standpoint, ESD remains a sectoral rather than a crosscutting policy agenda. Within government, the NSESD tends to be perceived as an environmental agenda. Accordingly, the process is not taken forward vigorously by most other agencies.

In particular, there is a strong political commitment to keeping ESD issues out of the micro-economic reforms that are now underway in Australia. This perspective is bound up with the geopolitics of federalism, in the sense that the states and territories perceive the NSESD as a 'Trojan Horse' that leads to a greater role for the Commonwealth government in environmental affairs. Furthermore, 'environment' is still widely regarded as a regulatory imposition, while the economic agenda is moving toward deregulation. Because of this attitude, NGOs and the informed public tend to consider that the NSESD process has lost momentum, and see it as a 'black hole'. Critics argue that the NSESD has had no perceptible influence and that 'business as usual' approaches persist (see also section above on NGO, public and political reactions).

Conflict Resolution

Current mechanisms for dealing with conflict and different opinions are largely internal to government and typically involve the usual bargaining and concession trading. As a result, the public at large and NGOs find it hard to gain a purchase on this process.

Public awareness is a vital prerequisite to effective involvement. In this regard, the post-UNCED and Agenda 21 process appears to be easier for professional associations, interest groups, local authorities and others to engage in. The now disbanded Resource Assessment Commission (RAC) provided an important means of giving strategic expression to ESD principles. The RAC had a high profile, and tackled some contentious resource use issues and conflicts (for example, timber inquiry). However, its reports were perceived as unhelpful to government decision-making and trade-offs.

Consensus

There is consensus across government and society on the concept of ESD and what is written in the NSESD — recognition of the need to do things in a different way, to integrate environment, social and economic considerations, and to work with all sectors. But 'beyond the words', there are many interpretations of what they mean and definitions are still very broad. Looking to the future, the way towards reactivating the NSESD is through NGO green groups, which still wield considerable political influence. 'Green preferences' were critical in electing Australia's last government, but environmental NGOs are now beginning to ask 'where the ESD strategy has gone' and this, in turn, may lead to more intensive lobbying and advocacy in support of the process. The danger here is that such a one-sided approach could reinforce the perception that NSESD is a green issue and thus further alienate industry and other pro-development interests (see sections above on key problems; key issues; and conflict resolution).

Relationship of NSESD to National Planning and Decision-Making

Integrated decision-making is a key principle of both the NSESD and the Intergovernmental Agreement on the Environment (IGAE). However, the whole idea of institutionalizing ESD has met with opposition. This is due, in part, to the 'hard slog' of developing the strategy in the first place. But the process also lacks continuity because central agencies did not want to pick up the implementation of the NSESD. As such, the relationship of the NSESD process to integrated national planning and decision-making is called into question. However, other key strategies and programmes do focus on sector-specific issues and problems, and these lend themselves more readily to conventional policy-making and planning.

The NSESD, like the IGAE, is not a legally-binding document. With the exception of legislation to establish the National Environmental Protection Agency (NEPA) in accordance with the IGAE, the strategy does not call for any legislation to be developed to make ESD operational. The name NEPA was changed to National Environmental Protection Council (NEPC) at the February 1994 meeting of the Council of Australian Governments and a bill to establish it was passed by the Commonwealth Parliament in October 1994. All state and territory governments have agreed (Western Australia, only belatedly, in late 1995) to introduce mirror legislation.

State and Local Strategies

Each state/territory has independently developed or is preparing 'subsidiary' strategies (some environmental, some dealing with sustainable development, some both). Since all states and territories have endorsed the NSESD, they take account of it. For example, South Australia is in the process of finalizing a state conservation strategy (its equivalent to an ESD strategy in relation to the state's living resources) and has completed a state water plan (a strategy for the ecologically sustainable use and management of the state's water resources).

In general, Agenda 21 has a low profile at local government level, but some states and territories are promoting Local Agenda 21s. In South Australia, a government officer has been appointed to assist local authorities in developing their own Local Agenda 21 plans, incorporated within their regular planning.

The federal government established a Local Agenda 21 project, which partly included the preparation of information kits to assist local councils implement Agenda 21. The project was completed in April 1994 and resulted in a guide and a small education campaign (DEST, 1994).

Driving Discipline

Sustainable development is the central focus of the NSESD but, as noted in the section above on key problems, this imperative is derived from a number of internal and external forces and, critically, these are now being overtaken by a micro-economic reform agenda, which is deliberately distanced from the NSESD.

Ecological Footprints and Transboundary Issues

Carrying capacity is not addressed systematically (for example, in terms of thresholds and limits) by the NSESD. However, environmental bottom lines, such as improving sink capacity in the national greenhouse response strategy, are referred to directly or implicitly. As an island continent, Australia has fewer and less immediately pressing trans-boundary issues than most countries — apart from French nuclear testing in the Pacific. There is a strong focus on global issues, including making environment and trade mutually supportive (for example, through GATT). This was a theme of Australia's 1994 report to the UN Commission for Sustainable Development (DEST, 1994).

Parliamentary Process

The NSESD was endorsed at the Council of Australian Governments' meeting in Perth in December 1992 by all heads of government. It is a policy document, but has no legal standing. It was approved by federal and state/territory cabinets. The Commonwealth Parliament passed the National Environmental Protection Council Act (1994) — a proposal of the NSESD. All state and territory governments have agreed to introduce mirror legislation (see section above on relationship of NSESD to national planning and decision-making).

Greening the Mainstream

Australia has made and is continuing to make a major push towards environment–economic integration, as described in Australia's 1994 report to the UNCSD (DEST, 1994). There is evidence (see Boxes 9.3 and 9.4) that, to a certain degree, there has been a significant 'greening' of industry. The NSESD has a major sector-specific thrust. On the other hand, micro-economic reform is still treated separately, and environmental 'bottom lines' are not treated systematically.

Press Coverage

There was much press coverage when the NSESD process was initiated, but much less when the strategy was published. Environmental/green matters are widely featured in the media, but are usually given sector and issue-specific treatment rather than an ESD orientation.

Review and Monitoring of the NSESD

In a personal communication, Paul Garrett of DEST writes:

> *The Australian Intergovernmental Committee for Ecologically Sustainable Development (ICESD) is coordinating a review of the NSESD. It has received information on the actions being undertaken by all levels of government to promote ESD within their particular jurisdictions. A range of legislative and non-legislative initiatives have either been, or are being, implemented across the eight sectoral and 22 cross-sectoral areas at which measures described in the NSESD are primarily targeted. Examples of specific actions include:*

- *incorporation of ESD principles into state and territory legislation. For instance, the state of New South Wales has passed legislation requiring the incorporation of ESD principles into the corporate objectives of commercialized government business authorities such as water and electricity authorities;*

- *the use of State Environmental Planning Policies (SEPPs) by the state of Tasmania as an environmental planning instrument. SEPPs address matters of statewide significance and which require a statewide application of policy. SEPPs can act as planning laws in their own right or can provide a framework for detailed planning at other levels; and*
- *the declaration of particular state government policies as 'sustainable development policies' where the state policy is found to comply with specific sustainable development objectives defined under a wider resource management and planning system.*

Apart from the various initiatives being pursued by each of the respective jurisdictions, the adoption and practice of ESD principles is also being pursued through the development of cooperative national initiatives to address environmental issues. Examples include the National Landcare Programme, the Commonwealth Coastal Policy (particularly the Coastcare component), establishment of the National Environmental Protection Council, the Rural Partnership Programme, and the formulation of national strategies dealing with issues such as the conservation of Australia's biological diversity, ecotourism and rangelands management.

Importantly, part of the current review process entails consideration of how future reviews of the NSESD might be conducted to ensure a more strategic assessment of its implementation and how this might best be achieved.

Chapter 10 CANADA

There have been two main national green planning and strategy processes in Canada. First, the *Green Plan* prepared by the government of Canada (led by the Department of the Environment, Environment Canada). It was first mooted in the summer of 1989 and, following 18 months of preparation, was released in December 1990 (Government of Canada, 1990a). It focuses on environmental action in areas under federal responsibility. The plan covered an implementation programme over six years (1990–6), but many targets had a ten-year perspective.

Second, the *Projet de société*, coordinated by the National Round Table on the Environment and the Economy (NRTEE), was initiated in November 1992 as a Canadian response to UNCED and Agenda 21. It is concerned with sustainable development issues at a national (not just federal) level. A strategy document, *Canadian Choices for Transitions to Sustainability* was published in June 1995. It is not intended as a 'static document or a submission to a particular level of government', but as a 'discussion paper for stakeholders designed to help them better coordinate their actions and see the larger picture, i e how sectors and regions interrelate' (Projet de société, 1995b). Thus, it has no set time perspective.

Most of Canada's provinces and territories have developed sustainable development or conservation strategies and environmental action plans, as have many municipalities, but these have not been linked directly to the Green Plan process or the Projet de société.

As pointed out in Box 6.6 (p55 above), the Department of Environment in Prince Edward Island was the first to develop a provincial conservation strategy in 1987. Alberta and Quebec also began working on conservation strategies in the mid-1980s. Yukon then released its own one in 1990. Most provincial sustainable development strategies are the product of provincial round tables. As mentioned above (p55), British Columbia, Manitoba, New Brunswick, Ontario, Nova Scotia and Saskatchewan have already prepared strategies. In Alberta the round table process enabled the identification of key areas for priority action and the development of a 'vision statement', as well as a statement about sustainable development principles. Quebec will hold an eco-summit in the autumn of 1966 with activities in the regions and, as also mentioned earlier, the Northwest Territories has implemented its own sustainable development policy. Various industrial and chemical companies, including Shell Canada, have also produced sustainable development strategies.

A number of sustainable development strategies have been prepared by local or regional municipalities (for example, Vancouver City Plan — a participatory planning process launched in late 1992; and Hamilton-Wentworth Vision 2020 — a sustainable development strategy for that region). Since 1987, local round tables have been appointed by city councils and others have 'emerged' and functioned at the provincial level (for example, in British Columbia, Manitoba and Ontario).

As shown in Box 6.6 (p55), a number of important regional initiatives exist:

- the Circumpolar Arctic Environmental Protection Strategy, signed in 1991, as an outcome of cooperation among eight circumpolar countries and indigenous peoples;
- the Arctic environment strategy (a component of the Green Plan);
- various Inuit-led initiatives, including the Inuit Regional Conservation Strategy which, as mentioned earlier, the Inuit Circumpolar Conference prepared as a response to the World Conservation Strategy;
- the work being undertaken to restore and protect the quality of the water and ecosystems of the Great Lakes and St Lawrence river by the International Joint Commission, which was established by the Boundary Waters Treaty of 1909;
- the attempts being made by the Atlantic Canada Opportunities Agency (ACOA) to introduce sustainable economic development to Nova Scotia and Prince Edward Island (see p55 above); and
- the Fraser Basin Management Program, the Gulf of Maine (marine) Action Plan, and other such initiatives for rivers and coastal areas.

There are various other sustainable development initiatives and processes in Canada that complement the national ones, for example:

- the Canadian Healthy Communities Project (1989–92);
- in the Great Lakes region, 17 communities have drawn up remedial action plans to revive and preserve waterfront areas such as Hamilton harbour;
- the Atlantic Coastal Action Plan to develop strategies or 'blueprints' for managing coastal resources of 13 communities in Atlantic Canada (this grew out of a Green Plan commitment to implement a marine environment programme);
- the so far eight self-help sustainability planning processes for small towns and rural communities set up since 1991 under Mount Allison University's Rural and Small Town Research and Studies Programme;
- watershed-based management plans and projects (for example, 75 community-based watershed management projects on Prince Edward Island, watershed advisory groups in Nova Scotia); and
- various programmes and projects aimed at 'greening communities' (focusing mainly on water and energy efficiency and conservation, water quality, waste reduction and management, greenspace planning, parks, natural areas, and wild-life habitat conservation).

The Canadian Biodiversity Strategy has been developed cooperatively with input from the federal, provincial and territorial governments and a non-government advisory group. Once it is approved, each jurisdiction will be responsible for implementing it within its priorities and fiscal capabilities. A draft was distributed for stakeholder review in June 1994 and a revised working group report was presented to environment ministers in November 1994.

In February 1994, Canada published its national report on climate change, identifying actions to be taken to meet commitments under the Climate Change Convention. The report was developed with the involvement of a multi-stakeholder group, including federal departments, provinces, municipal governments, and stakeholders in the environmental and business communities. A national Climate Change Action Program was developed under the guidance of the National Air Issues Coordinating Committee to propose measures to reduce domestic greenhouse gas emissions. Discussions with key stakeholders began in the summer of 1994 and were broadened during the year. The programme was presented to the first conference of the Parties to the Convention in Berlin in March 1995. Although both the Biodiversity Strategy and the Climate Change Action Program have been multi-stakeholder processes, there has been no direct linkage with the Projet de société.

The Green Plan

Prime Motivation and Getting Going

Political demand for the Green Plan grew out of increased environmental awareness in the late 1980s. Analysis of opinion polling shows that at first the public was concerned mainly about local air quality. Public concern was raised following accidents involving PCBs. With rising living standards, people came to value more their quality of life. Transboundary and international issues gradually assumed greater importance.

A Conservative government was in power between 1984 and 1993. The Minister of Environment lost public credibility because of budget reductions and there were problems over jurisdiction dilemmas between the federal and provincial governments. But in the late 1980s, in the run-up to UNCED, 'the environment' became the number one public issue and there was a demand on the federal government 'to do something'. The definition of 'what was to be done' was left to federal bureaucrats.

After publication of the Brundtland Commission's report (WCED, 1987), the government set up a National Task Force on the Environment and the Economy. Its recommendations led eventually to the establishment of the National Round Table on the Environment and the Economy.

During the period 1986–8, the government went through 'an open and transparent process' that led to proclaiming a new Canadian Environmental Protection Act (1988), which consolidated and expanded environmental protection legislation. The government also prepared its response to the World Conservation Strategy.

There were several further important changes in the late 1980s: Lucien Bouchard became the Minister of the Environment, a new Deputy Minister was appointed, a Cabinet Committee on the Environment was established, a new policy group was created within Environment Canada, and there was an increase in Environment Canada's budget (mainly to fund issue-specific responses, such as action on the Great Lakes and St Lawrence). These institutional changes had an influence on the Green Plan process, which was initiated in the summer of 1989. It was based on a 'more pragmatic approach to environmental management within the context of sustainable development'. As a result, the fiscal aspects of the Green Plan dominated.

Although there was government-wide commitment to the Green Plan process, with Environment Canada taking the lead as catalyst and coordinator, interdepartmental relationships became clouded by fiscal issues. A relatively large proportion of 'new' government money was made potentially available to implement the Green Plan, and at a time of more general financial restraint and retrenchment. Environment Canada thus found itself evolving into a major department (with a significant budget). The minister (M Bouchard) succeeded in linking the Green Plan to the federal budget process. It therefore became subject to Cabinet 'secrecy'. Internal government documents on the Green Plan were marked 'secret' and there were strict limitations on who could see them. An early copy of a draft Green Plan was 'leaked' to the 'environment community'.

Focus

The Green Plan is 'mainly an environmental strategy in a sustainable development context'. The focus is on sustainable decision-making — very much economic decision-making in relation to environmental problems; and on an action plan to address specific issues (such as climate and fisheries).

In considering a follow-up to the Green Plan, the government is seeking better to integrate sustainable development — not just environmental considerations — into all federal government decisions on programmes, policies and operations (see section below on the follow-up to the Green Plan).

Lessons about the environment–economy intersection in Canada can be drawn from the report of the Task Force on Economic Instruments and Disincentives to Sound Environmental Practices (1994), which was set up by the Minister of Finance and the Deputy Prime Minister. The current Liberal government's election 'red book' (Canadian Liberal Party, 1993) also provides guidance on its environmental thinking, which is likewise guiding the Green Plan follow-up process.

Organization and Management

The Green Plan was coordinated and prepared by a small policy group in Environment Canada. A small group also managed the consultation process (see section below on participation) and workshops. Most of the writing was done by a drafting team, but through an iterative process. Some expertise was 'bought in' to deal with editing, translation, report production and communications. The Environment Canada 'team' prepared initial sector drafts, which the sectoral departments then reviewed and revised. The team edited these versions and negotiated changes.

A few members of the Green Plan team had been involved in the preparation of the national Energy Policy. They also took note of the experience of the Dutch government in preparing its *National Environmental Policy Plan* (VROM, 1989) (but only when the Green Plan was well advanced).

Terms of Reference

There were no set terms of reference. Environment Canada was instructed by the Minister of the Environment to develop Canada's position on environmental policy within an initial period of six months and had to interpret government thinking on this issue. A detailed report was prepared and submitted to the Cabinet and, following this, the latter decided to undertake extensive consultations. To allow time for these, the period for delivery of the Green Plan was then extended to 18 months. Environmental groups had pushed for more consultations. M Bouchard also resigned and this caused an interruption in the process — the new minister took time to 'come up to speed'.

A budget of C$ 3 billion was established for a five-year Green Plan. This was subsequently 'diluted': first the period was extended (in the 1991 budget) to six years; then reduced to $2.5 billion over two successive subsequent budgets (1992 and 1993).

In reviewing the Dutch experience (see Chapter 14), the Green Plan team considered the Dutch goals, objectives and targets too detailed and too specific for Canada — although some targets were subsequently introduced. The model for developing the Green Plan, described in the booklet *A Framework for Discussion on the Environment* (Environment Canada, 1990), was mostly defined through in-house discussions.

Participation

Though the government made a deliberate decision to 'own' the Green Plan process and its products, rather than be a 'partner' along with others in a wider process, it did, however, aim to 'involve' others. This decision was taken at a time when the 'public was calling for "leadership" from the federal government'.

Ministers of the main departments were involved as a committee throughout. In addition, a Green Plan coordinating committee (mainly for implementation) was established in which other government departments participated. 'Environment Canada was successful in getting other departments to participate, but there was "money on the table". To some extent, the Green Plan was the only "game in town". The follow-up to the Green Plan will not be a money document.' The first draft was sent to Cabinet about half way through the process (after about nine months). But there was a feeling in Cabinet that there had been insufficient consultation. This coincided with a growing 'anti-feeling' in the private sector. The Cabinet authorized consultation and some of the 'secrecy' was lifted. The approach to the consultation process is discussed by Environment Canada (1990).

A large 'consultation' process followed, which is described in a government report (see Government of Canada, 1990b). Environment Canada estimates that 10,000 people participated in the process in one way or another — attending meetings, writing letters, or preparing briefs. Its aim was to strike a balance between leadership of the process and public participation.

NGO, Public and Political Reactions

A number of NGOs were negative — although some 'were confrontational and might have been negative whatever the product'. However, many conservation-oriented NGOs were pleased with the resources being channelled into nature conservation.

Environment Canada believes that, with the publication of the Green Plan, public appreciation of the government's performance on the environment increased. Its performance and 'leadership' in this area were viewed as its best achievements at the time — but this was within a context of being unpopular on all other fronts and the government was severely defeated in the 1993 elections. In defining its party platform for the elections, the Liberal Party set a clear and positive position on the environment in its Red Book. The Green Plan was seen as 'politically owned by the Conservative Government'. When the new Liberal government took office, it remained relatively silent on the Green Plan, but its Red Book provided a clear and positive steer on its commitment to environmental issues (Canadian Liberal Party, 1993).

Key 'Assistance' Factors, Problems and Issues

Two important factors provided impetus for the launch of the Green Plan: public pressure for action on the environment, and government commitment at the highest level. But there was a series of problems during its development and implementation:

- It was difficult to design an appropriate process that recognized that ministers 'have to decide'. Conditions have now changed and it is now easier to design an open Green Plan follow-up process leading to ministers taking final decisions.
- The number of staff dealing with public policy is relatively small within government and there are a lot demands placed on them. Thus, a strong central directive is required to mobilize this 'talent' towards any individual initiative. Getting individuals to focus and contribute was a problem. Adding 'consultation' increased the complexity of this problem across government.
- It was difficult to secure resources to implement the Green Plan. As it neared completion, it was unclear whether it would receive final approval. There were constitutional problems over Quebec, interest in the environment had peaked, and the government focus was shifting as the financial situation deteriorated with economic decline. As one government officer commented,

  > *The Government has shifted its focus to creating economic opportunity and a 'jobs and growth' agenda with virtually no new resources for public policy. Current advancement of the sustainable development agenda focuses on efficiency and effectiveness — getting the institutions right — in a milieu of fiscal restraint.*

- The Green Plan was released in December 1990, but in the government budget announced the following February, its budget was diluted by spreading it over six rather than five years. Stakeholder groups doubted the government's seriousness and commitment. Government signals were therefore unclear.

There was an early decision to make the Green Plan comprehensive. Pressure to set priority issues was rejected.

Conflict Resolution and Consensus

A combination of mechanisms was established to develop the Green Plan. These included consensus-building within and between government departments to the fullest extent possible. There were several levels of decision-making with ministers eventually resolving difficult issues. But inevitably, in the end, there had to be one decision-taker — Environment Canada. Technical documents were circulated regularly. There were parallel financial negotiations on budget allocations.

Environment Canada asked all sectors 'to be creative on what needs to be done'. Its role was seen as synthesizing, shaping and ensuring that the Green Plan 'lived within bounds' (i e within available financial resources).

The Cabinet Committee on the Environment was dedicated to dealing with the Green Plan process (in addition to other issues), taking high level conflict-resolving decisions on differing opinions within government (particularly concerning financial or regulatory issues and on issues concerning doing things in-house versus externally).

The summer of 1990 was a 'reconciliation period' when the results of the consultation process were analysed. According to Environment Canada, 80 per cent of all issues raised publicly are addressed (not necessarily to everyone's satisfaction) in the Green Plan. An Environment Canada officer commented that 'The consultations were helpful in affirming some of the directions that we were moving in, and in highlighting further areas of concern and opportunities.' The consultations also 'reaffirmed the importance of moving ahead on the Green Plan', and had some influence on it. In particular, comments on wildlife, agriculture and forests materially assisted the design of those chapters of the Green Plan.

Environment Canada staff interviewed stated that there was consensus about the Green Plan within government. Within society as a whole, there were criticisms (some from 'professional critics') — perhaps because expectations may not have been fulfilled, but the government was credited with having produced a plan and implementing it. While NGOs were mostly critical about the overall plan, there was support for individual components.

Links to National Planning and Decision-Making, Regional Strategies and Convention-Related Strategies

The Green Plan was closely integrated with the budgetary process. It had built-in targets and schedules — as mechanisms for public accountability. Politicians were able to be held publicly accountable. The Green Plan was a policy document. It did not itself have the force of law and was not binding. However, it did contain international commitments and many Green Plan initiatives did have the force of law, for example, some legislative changes were introduced.

The Green Plan was not binding on provincial or municipal governments, and federal-provincial agreements were used to implement its provisions (for example, in agriculture). It was closely linked to various regional strategies:

- According to the Environment Canada staff, the North American Free Trade Agreement (NAFTA) — North American Agreement on Environmental Cooperation (final draft, 13 September 1993) — would probably not have been concluded without the Green Plan.
- Canada's implementation of the 1991 Arctic Environmental Protection Strategy (AEPS) was funded under the Green Plan (C$ 100 million). The AEPS had its origins in the Circumpolar Conservation Strategy — originally called the Finnish Initiative and developed by agreement between eight nations.
- The Green Plan was the basis for Canada's negotiations at UNCED. The process to develop Canada's UNCED national report was considerably more open than the Green Plan.
- The Canada–USA Great Lakes Water Quality Agreements (GLWQA) preceded the

Green Plan. But the Green Plan became a vehicle for expanding and extending actions to meet the commitments under the GLWQA.

The Green Plan established Canada's negotiating position on the Conventions on Biodiversity and Climate at UNCED. A report of the biodiversity working group (November 1994) was prepared in response to the UN Convention on Biological Diversity. Environment Canada leads within government on biodiversity, but on climate, leadership is shared jointly with Natural Resources Canada (which is responsible for energy).

Driving Discipline, Ecological Footprints and Transboundary Issues

According to Environment Canada staff interviewed,

> *intellectual thinking was done by economists, it was 'stick handled' by them. The central paradigm was that environmental problems are decision-making problems and therefore we have to assimilate environmental problems into economic decision-making. We have to internalize the environment.*

The Green Plan did not specifically address the environmental impact of Canada's trade patterns. But the 'issue of "ecological footprints" is built into Canada exercising its stewardship responsibility. There has been much discussion of Canada's role in the world. However, most of the key environmental issues in Canada are international and transboundary ones'. The Green Plan discussed and committed Canada to working towards accord with the USA on SO_2. The Canada–USA Air Quality Agreement brought about the updating of the US Clean Air Act (1991/2). 'The Green Plan embodied what was going on anyway, but laid out a game plan.'

Parliamentary Process

The Green Plan comprised 80 individual programme elements, each of which was reviewed by the Cabinet Committee on the Environment and by a treasury board process. This review process took about one year. The plan was tabled in Parliament and there was a debate on its general contents (but no votes). There were, however, debates and votes on legislative changes required as a consequence of elements of the Green Plan.

The Follow-Up to the Green Plan

The current federal government is promoting sustainable development as an integral component of decision-making at all levels of society — an approach outlined in the Liberal Party's election Red Book (Canadian Liberal Party, 1993). The aim is to 'integrate environmental and economic considerations, and the social aspects of sustainable development into government decision-making as our understanding of these aspects grows'. Thus, at the federal level, the need for sustainable development is now being internalized within policy development and decision-making processes by individual sectoral departments and agencies and Crown corporations. The Departments of Finance and Environment also established a multi-stakeholder task force to find effective ways in which to use economic instruments to protect the environment and to identify barriers and disincentives to sound environmental practices. Its report, published in November 1994, was designed to influence the 1995 budget.

A presentation by Environment Canada staff was made to the department's management committee on how to deal with the government's commitment that 'environment and economic signals should point the same way'. Before making the presentation, Environment Canada staff talked with other federal government departments and largely agreed on a position. This was endorsed by the management committee. An interdepartmental committee was established to steer the follow-up process.

During 1995 there was a government-wide review of programmes. All federal departments reviewed their spending within the context of what was necessary, and what was affordable. As part of this review, Green Plan programmes were integrated into each department's activities so that they are now no longer identified separately.

Legislation has been passed by Parliament to establish a 'Commissioner of the Environment and Sustainable Development' to hold the government accountable for 'greening' its policies, operations and programmes. The legislation also requires all federal ministers to table departmental sustainable development strategies in Parliament. These strategies should be comprehensive, results-oriented, and prepared in consultation with partners, clients and stakeholders. In June 1995, the government released *A Guide to Green Government* (Government of Canada, 1995), signed by all Cabinet Ministers and representing a government-wide commitment to sustainable development, which provides a framework to assist in the preparation of departmental strategies.

> *The Guide will assist departments in meeting their obligations, under amendments to the Auditor General Act tabled in Parliament in April [1995], to deliver sustainable development strategies with concrete objectives and action plans. These strategies will have to be completed and tabled in Parliament within two years of the coming into force of the amendments. A new Commissioner of the Environment and Sustainable Development will monitor and report annually to Parliament on departments' progress in implementing their action plans and meeting their sustainable development objectives.*
>
> (Environment Canada press release, 28 June 1995)

The three-part guide (see Box 10.1) was developed with the assistance of an advisory group with the following composition: business (3), environment (3), academic (3) and federal government departments (3) — from Industry, Natural Resources and Environment Canada), and an independent chairperson — an academic with some experience as an independent adviser to the Deputy Minister of the Environment during the green planning process. The guide complements the proclamation of the Canadian Environmental Assessment Act, which ensures that the environment is taken into account in the planning of projects.

Box 10.1 CANADA'S GUIDE TO GREEN GOVERNMENT

The Canadian government's A Guide to Green Government *is designed to assist all federal departments in preparing sustainable development strategies. It comprises three parts:*

I. ***The Sustainable Development Challenge*** *translates the concept of sustainable development into terms that are meaningful to Canadians, underscoring its important social, economic and environmental dimensions. A series of sustainable development objectives are presented (for example, using renewable resources sustainably, preventing pollution, fostering improved productivity through environmental efficiency) that represent a starting point for the preparation of departmental strategies.*

II. ***Planning and Decision-Making for Sustainable Development*** *sets out the policy operational and management tools that will facilitate the shift to sustainable development. It encourages an integrated approach to planning and decision-making that considers social, economic and environmental factors that play a role in providing a high quality of life, based on the best available science and analysis, and visions and expectations of Canadians. The approaches discussed include: promoting integration through the use of tools such as full-cost accounting, environmental assessment and ecosystem management; developing strategies by working with individuals, the private sector, other governments and Aboriginal people; using a mix of policy tools such as voluntary approaches, information and awareness tools, economic instruments, direct government expenditure, and command and control.*

III. ***Preparing a Sustainable Development Strategy*** *presents the main elements that departments could consider as the basis of their strategies. It is recommended that 'strategies should all be results-oriented, showing in clear, concrete terms what departments will accomplish on the environment and sustainable development; comprehensive, covering all of a department's activities; and prepared in consultation with clients and stakeholders'. Six steps are suggested for the preparation of strategies:*

- *Preparation of a* ***Departmental Profile****, identifying what the department does and how it does it.*
- ***Issues Scan:*** *assessment of the department's activities in terms of their impact on sustainable development.*
- ***Consultations*** *on the perspective of clients, partners and other stakeholders on departmental priorities for sustainable development and how to achieve them. It is suggested that a brief report 'on the nature of the consultations and how views contributed to the final product would be useful for partners and stakeholders, and contribute to openness and transparency in the preparation of strategies'.*
- *Identification of the department's* ***goals and objectives and targets*** *for sustainable development, including benchmarks it will use for measuring performance.*
- *Development of an* ***Action Plan*** *that will translate the department's sustainable development targets into measurable results, including specifying policy, program, legislative, regulatory and operational changes. 'Because sustainable development is a shared responsibility among departments, governments, Aboriginal*

people and other stakeholders, implementation of action plans will likely require cooperation and partnership. In these instances, departmental strategies should describe the cooperative mechanisms and partnerships that will help them achieve the targets, objectives and, eventually, their goals'.

- *Creation of mechanisms to monitor (measurement and analysis), report on and improve the department's performance.*

Source: Government of Canada, 1995.

The Projet de Société

Prime Motivation and Getting Going

The origins of the *Projet de société* can be traced to Canada's participation at UNCED, where a broad multi-stakeholder delegation adopted a 'team Canada' approach. There was pressure to keep this going after Rio and discussions — facilitated by the National Round Table on the Environment and the Economy (NRTEE) — were held to consider how Canada should maintain the UNCED momentum and process. The idea of adopting a multi-stakeholder process to develop a sustainable development strategy for Canada arose from these discussions.

Representatives from more than 40 sectors of Canadian society, including business associations, community organizations and indigenous peoples, attended the first meeting of the National Stakeholders' Assembly in November 1992. Here they agreed to accept the then Environment Minister Jean Charest's bid — in a speech in Parliament in November 1992 — to respond to the commitments of Rio and the challenge of sustainable development through a '*Projet de société*', which, in his own words, 'embraces society as a whole and aims at becoming a driving force, a factor transcending our usual limits'. The National Stakeholders Assembly was not a representative assembly, but a coalition acting as a network of networks held together to work towards achieving tasks that could not be achieved separately.

Each of five 'sponsoring' organizations (the Canadian Council of Ministers of the Environment (CCME), Environment Canada, the International Institute for Sustainable Development (IISD), the International Development Research Centre (IDRC), and the NRTEE) contributed C$ 50,000 to establish a secretariat and hire a research director. A working group was established with two subcommittees, one called 'documentation and information', and the other 'vision and process'. These assumed responsibility, respectively, to analyse Canadian responses to Rio and to draft a concept paper on sustainability planning.

The NRTEE facilitated and chaired the process and provided the secretariat. Most of the tasks were undertaken by volunteers and committees, which met monthly. But there was tension between those who wanted 'to develop strategic plans' and those who wanted 'to do specific projects'. It was decided to do both. A progress report and recommendations were presented to a second National Stakeholders Assembly in June 1993. Politically, this was an unfortunate time because M Charest was involved in a party leadership campaign and was unable to attend. Thus, there was no senior Conservative at the meeting.

By the time the third assembly met in December 1993, the political environment had changed. A new Liberal government had assumed power following a landslide election victory. The new Environment Minister (also Deputy Prime Minister) was supportive of the *Projet de société*, but there was no money available for the process from government, which became heavily concerned with budget deficit reduction. The assembly asked the NRTEE to assume a larger management role, rather than merely act as a facilitator, for the next phase of the *Projet de société*. It was also evident that many volunteers were becoming 'tired' and that there was a need 'to move faster and produce a draft strategy'. The NRTEE therefore hired two policy advisers to facilitate the preparation of a draft strategy document. These advisers worked closely with a volunteer working group which met in March, May and September 1994 to criticize and revise the evolving document which was tabled at the fourth assembly in November 1994 (this time funded by the NRTEE) as *Canadian Choices for Transitions to Sustainability* (Projet de société, 1995b).

The first day of the meeting included concurrent workshops on existing sustainable development activities in Canada: 'municipal sustainability planning, sectoral sustainability initiatives, provincial sustainable development strategies, international strategies and Aboriginal peoples, federal and national strategies, economic instruments, and linking Canadian strategies to international concerns'.

The second day focused on the work of the *Projet de société*. The draft strategy document was considered. Many minor changes were suggested and some major concerns were raised, but the document was endorsed. A revised document was published in January 1995. The NRTEE organized a series of about 12 meetings across the country to determine how useful such a document might be in engaging various constituencies in discussions about sustainability. These meetings focused for the most part on the choicework tables (see Box 10.2 and Table 10.1) and the transition tools, although they were also used as an opportunity to elaborate on the introductory and concluding chapters, and to update the description of Canadian sustainability initiatives.

The working group was reconstituted in early 1995 and completed the strategy document, based on the feedback received. The strategy was published in June 1995. The working group also compiled a directory of sustainability tool kits for communities. Sustainable livelihoods was selected as a focus for the working group activities with a forum on this subject planned for March 1996.

Focus

The work of the *Projet de société* was process-oriented and aimed to engage all interests, communities and constituencies in considering how to work towards and become involved in initiatives that promote sustainable development (see goals in Box 10.3). An immediate goal was to design a strategy that built on the many relevant activities already underway in Canada (see section above on strategy processes in Canada) by governments, business and industrial associations, and local communities, which would give them a clearer strategic focus and linkage.

Box 10.2 CHOICEWORK

To assist stakeholders to reach innovative solutions, the strategy of the Canadian Projet de société (June 1995) attempts to reduce the 'blinders' of sectoral bias and traditional mandates by providing innovative choicework tables around basic human needs, such as air, water, food and mobility. An example of a choicework table is provided in Table 10.1. Chapter 4 of the strategy, 'Canadian Choices for Transitions to Sustainability', defines 'choicework' as 'sorting out choices, weighing pros and cons, and beginning making the difficult trade-offs'. The tables attempt to 'compare expert and public perceptions of various issues in order to find a method to bridge the gap between experts and the general public on a range of sustainability issues'. The tables also identify areas of conflict and levels of consensus in order to show where immediate progress can be made and where more consensus-building is needed.

Organization and Management

From 1992, a secretariat for the *Projet de société* was provided by the NRTEE. The Projet de société was an 'organic process — a virtual organization'. It had no official office or budget as such — it was facilitated. It was a new approach, which 'lay in the interstices of government and non-government'. Tasks were originally undertaken by volunteers working through committees, which met regularly, with progress reports being presented to the National Stakeholders Assembly. From late 1993, the NRTEE assumed more of a managerial role.

Early work has focused on three complementary and reinforcing activities:

- a report on Canadian responses to Agenda 21 and the Rio conventions;
- developing a framework for integrating the environmental, economic and social agendas, and a process for building consensus and coordinating efforts of all sectors toward a common purpose; and
- practical actions for moving forward.

In 1994, work intensity increased, and two NRTEE staff members helped to prepare a draft strategy document. They reported to a strategy working group which met three times before the fourth National Stakeholders Assembly, which reviewed the document and endorsed its publication (in January 1995) as a working tool.

A communications strategy was developed and a programme established to create affiliation for 'activities and initiatives that advance the sustainability movement'. In this, the Projet de société associated with the SustainABILITY programme, an idea launched by ParticipACTION and the NRTEE (see Box 10.4).

Terms of Reference

Terms of reference were set out in a 'Prospectus' (May 1994), which introduced the Projet de société and described its work as a multi-stakeholder coalition acting as a

Table 10.1 Example of Choicework Table

Mobility

Some examples of choices that could be considered	Timing Duration Impact	Costs: $, Environ., Social	Benefits: $, Environ., Social	Some consequences	Partnerships	Responsibilities	Consensus levels
Replace vehicle registration fees with "feebates": rebates for efficient vehicles; fees for inefficient vehicles	months years xx	$ ss	$ eeee sss	Would increase efficiencies and ensure that the polluter pays	Car dealers	P	?
Negotiate covenants with insurance industry to facilitate car pooling and sharing and pay-at-pump insurance	months years xx	$$ ss	$$ eee ssssss	Higher vehicle occupancy; more jobs in car leasing industry; fairer distribution of insurance costs	Commuters and insurance industry	F P B	?
Reduce the deficit through dedicated increases in excise taxes on fossil fuels	months decades xxxxxx	$ sss	$$$$$ eeeeee ssss	Would take advantage of concern over deficit to reduce CO_2 emissions and respect UNCED commitments	Public transport and car servicing industry	F	?

Legend

Timing: Time it would take to implement choice. **Duration:** Period during which the impact is felt. **Impact:** x = low impact; xxxxx = high impact. **Cost:** $ = low monetary cost; eee = medium environmental cost; ssssss = high social cost. **Benefits:** $$$$$$ = high monetary benefit; eee = medium environmental benefit; s = low social benefit. **Consensus:** ✓ = low consensus; ✓✓✓✓✓✓ = high consensus. **Responsibilities:** F = federal; P = provincial; M = municipal; B = business; C = civil society.

Source: Projet de société (May 1995)

Box 10.3 SUSTAINABILITY GOALS FOR CANADA

Goal 1: Air *Ensure that all Canadians have safe air to breathe at all times while maintaining socioeconomic activities that do not threaten global climate security.*

Goal 2: Fresh and Salt Water *Provide access to safe drinking water and economical supplies of water for other purposes without reducing the capacity and quality of water resources, including salt water resources.*

Goal 3: Food *Ensure a sustainable system of food production, distribution, processing, consumption and recycling that promotes healthy diets and strong economies, both at home and around the world.*

Goal 4: Habitat: Human and Natural *Provide adequate shelter and a sense of community for all Canadians without threatening the natural habitat upon which people and all other species depend.*

Goal 5: Human Relationships *Develop a society in which people respect one another for their differences and their common human values, and in which people are able to develop to reach their full potential, without compromising similar opportunities for future generations.*

Goal 6: Health *Create a healthy environment and an affordable health care system that will improve the physical and mental well-being of all Canadians.*

Goal 7: Security *Develop a community, a country and a world in which there is much less danger, fear and worry.*

Goal 8: Mobility *Ensure levels of mobility and communication that support basic human needs without denying future generations similar opportunities.*

Goal 9: Sources and Sinks *Use resources at a rate that ensures that the regenerative (source) and assimilative (sink) functions of natural ecological systems are maintained.*

Source: Projet de société, *Newsletter*, January 1995.

catalyst to 'help promote Canada's transition to a sustainable future', and as a forum for focusing on sustainable development issues and encouraging creative responses. It invited the participation of interested parties in future activities. This document set out the following principles and characteristics of the Projet de société:

- The process is designed to be transparent, inclusive, and accountable;
- Each partner and each sector is encouraged to identify and take responsibility for its own contribution to sustainability;
- Dialogue and cooperation among sectors and communities are key elements of problem-solving;

Box 10.4 SustainABILITY

With financial support from the federal government, all provincial and territorial governments, plus the J W McConnell Family Foundation, Nestlé Canada, the Communications, Energy and Paperwork Union of Canada, CIDA and IISD, the NRTEE and ParticipACTION have produced a comprehensive strategy for a national communications program in support of sustainable development. SustainABILITY is the result of a six month consultation process that engaged 180 diverse experts and leaders across Canada. Rather than duplicate existing efforts, it will be a supportive communications campaign that should reinforce, promote and enhance existing initiatives. It was presented to the federal government in December 1994, but has not yet been funded.

(*Projet de société*, Newsletter, *December 1994*)

Note: *ParticipACTION is a non-governmental agency promoting physical fitness in Canada through social marketing.*

- A shared vision and agreement on key policy, institutional and individual changes are necessary for the transition to sustainability;
- Strategy and action must be linked, and must build on previous and ongoing initiatives; and
- Canada's practice of sustainable development and its contribution to global sustainability should be exemplary.

Participation

The Projet de société was probably one of the most participative strategy processes in the world. It was 'a multi-stakeholder partnership of government, indigenous, business and voluntary organizations' which operated through collaboration and consensus-building. Representatives from more than 100 sectors of Canadian society participated in the National Stakeholders Assemblies. They are listed in the projet prospectus (Projet de société, 1994a).

NGO, Public and Political Reactions

As a multi-stakeholder process, First Nations, governments, NGOs, businesses and the public were all integrally involved in and supportive of it. There were extensive discussions and debates and the Projet de société itself was healthy and progressive, but it did not prove possible to build consensus across all interests represented.

Mandate

The *projet* had no official mandate, but one was defined by the original group of stakeholders, at the first National Stakeholders Assembly in November 1992, to review the commitments Canada had made at UNCED and to develop and pursue planning for a sustainable future. This was in response to the Minister of Environment's speech in Parliament in November 1992 challenging Canada to meet the commitments it had made at Rio. The process was recognized and supported by government and by all the 100 sectors that participated. It therefore had an unofficial but broad public mandate.

Key 'Assistance' Factors

A number of factors helped the work of the *Projet de société*. For example, the various provincial strategies (in most cases facilitated by provincial round tables) provided a good basis of experience. Many of those involved in these strategies also became involved in the *Projet de société*, injecting their experience, enthusiasm and support into it. The support provided by the NRTEE itself was crucial. To an extent, the 'bad' experience of NGOs, trade unions and others of the Green Plan, which they perceived as a 'top-down' process, encouraged their participation in the work of the *Projet de société*. Many of the individuals involved in it already had 10–20 years' experience in the field of sustainable development and a history of involvement in national multi-stakeholder processes (especially those in environmental NGOs). There was also a pre-existing 'collaboration culture' at the national level from, for example, the previous six years' experience of the now defunct Canadian Environmental Advisory Council, the Coalition of Energy, Environment and Development Groups that attended the UN Conference on Renewable Energy, and the round-table movement.

Key Problems and Issues

There were also a number of problems which hindered the *Projet de société*. For example, individual personalities had strongly-held points of view (both a key strength and a key problem). There were tensions between those wanting action and those preferring to focus on policy discussions, thus making it difficult to reach a consensus. Because the process was based on inclusiveness — anyone who represented a sector could become involved — every time a new stakeholder joined, past discussions and decisions had to be revisited, which slowed down the process. On many 'big issues' (such as climate change) there was (and is) no national consensus.

The traditional, institutionalized response of stakeholders was to 'protect their own community' or 'fight their own corner' and, even though they were willing to cooperate, they tended to defend their own interests first. Much effort was needed to maintain the business sector's interest and willingness to attend. Decisions made by companies tend to have a greater impact on sustainable development than many of those made by government. Numerous companies became involved in the *Projet de société* because they were concerned about what the government might do as a regulator, but when they realized that it was 'only' an independent, multi-stakeholder forum and not an agency of government, they became less keen to participate.

The work of the *Projet de société* had to take account of the fact that Canada is a very large country with strong regional interests, which makes the task of devising a national strategy that could remain relevant to the whole country particularly difficult. For example, forest issues dominate British Columbia, fisheries are more important in Newfoundland, while oil-rich Alberta is more interested in climate change and would not wish to see a reduction in the use of fossil fuels. While the debt and national unity crises were perceived to pertain to the whole of Canada, there was still no 'perceived' crisis as regards sustainable development, which then slipped down the 'agenda' of critical issues. Though Canada's fishing industry is of national importance, it was treated as a regional issue.

It was difficult to secure the deep involvement of Quebec stakeholders, even though the *Projet de société* was conducted in a bilingual and inclusive manner. It had to deal with being a comprehensive approach across a diverse and large country. It also had to accommodate topical issues, which tended to 'flare up' (for example, fisheries were a major issue at the second National Stakeholders Assembly, while indigenous peoples were a key feature at the second and third assemblies.

Conflict Resolution

Many difficulties were resolved. The *Projet de société* was a continuous process of acting as an 'honest broker' and the strategy addressed both the process and its products. A key feature was to celebrate the achieving of conflict resolution through tête-à-tête meetings, linking individuals together (brokerage on a personal level).

A few stakeholders left the project, particularly when their involvement was driven by a single issue, but others developed their own multi-stakeholder processes based on their experience within the *Projet de société*. The Mining Association of Canada, for example, set up an independent process known as the Whitehorse Mining Initiative (WMI). This brought together, for the first time, all legitimate mining stakeholders in Canada. It culminated, in September 1994, in the signing of the WMI Leadership Council Accord at the Mines Ministers' Conference in Victoria, British Columbia.

In developing a strategy for sustainable development, difficult choices and trade-offs had to be made, particularly where consensus could not be reached. To assist stakeholders to reach innovative solutions, the draft strategy document attempted to reduce the 'blinders' of sectoral bias and traditional mandates by providing innovative 'choicework'[1] tables around basic human needs, such as air, water and food. These tables attempted to 'compare expert and public perceptions of various issues in order to find a method to bridge the gap between experts and the general public on a range of sustainability issues'. The tables also identified areas of conflict and levels of consensus to show where immediate progress could be made and where more consensus-building was needed.

Ecological Footprints and Transboundary Issues

Although the idea of 'ecological footprints' was developed by a Canadian (William Rees), the strategy document developed by the Projet de société does not discuss this issue directly. However, the document recognizes that Canada is one the world's most internationally exposed countries (with the second largest landmass and borders on three oceans). In a section on 'international dimensions', Canada's environmental vulnerability and the idea of ecological footprints are introduced indirectly through a brief discussion of the effects of southern European fishing fleets on fish stocks in Canadian

---

1. Sorting our choices, weighing pros and cons, and beginning to make the difficult trade-offs is called 'choicework'.

waters, and through discussion of consumption patterns. Transboundary issues are alluded to, for example through mention of airborne nuclear waste drifting in from the former USSR. In dealing with such issues, Canada places considerable reliance on building strong multilateral institutions.

Press Coverage

There was practically no press coverage of the Projet de société's work or its strategy document. The general public was not very aware of the process.

Postscript

In a personal communication, Sandy Scott, former policy adviser to the Projet de société, writes:

> *The National Round Table on the Environment and the Economy (NRTEE) decided in the fall of 1995 to end its support for the Projet de société. While the Round Table was pleased with the success the Projet de société had had in developing its national strategy, Choices for Transition to Sustainability (Projet de société, 1995b), it did not see a role for itself in continuing to support this multi-stakeholder initiative.*
>
> *While this stage of the Projet de société has, as a result, come to a close, many of those who have participated as stakeholders in the process hope that a new process will one day emerge to take up where the Projet de société left off. The latter was, from its beginning, an experiment in new ways of governing. It may be that our hopes of being an institution without walls, or a 'virtual organization', meant that, in the end, it was easier for decision-makers to abandon the initiative in a time of fiscal constraint and changing political priorities.*
>
> *As its final commitment, the NRTEE did agree to finance a workshop to be held in March 1996 to explore the key policy questions around jobs and sustainable livelihoods, an issue that was emphasized throughout the Canadian Choices document. Projet de société members hope that the forum,* Working Toward Our Common Future, *will continue to show the value of working in multi-stakeholder forums and of trying truly to address the environmental, social and economic challenges we face in an integrated way. Though people and names will change, the lessons learned through the Projet de société will stay with us. Undoubtedly this has been a stepping stone in the ongoing process of figuring out how to DO sustainable development!*

Chapter 11 DENMARK

The Danish Ministry of the Environment and Energy (MoE) was one of the earliest environment ministries to be established (in 1971) and, periodically, it has prepared comprehensive environmental statistics. Over the last ten years, a variety of strategic plans have been developed by individual ministries — some broader than others — as part of a rolling process, for example, for transport, energy and agriculture. The MoE plays an umbrella role, assisting concerned ministries to prepare and review these plans. In 1995–6, the energy and agriculture plans have been reviewed.

Foundations for Strategic Environmental Planning

In Denmark, there is a good foundation for policy-planning based on excellent environmental data, statistical reports and regular state of the environment reports. The report, *Environment and Development: The Danish Government's Action Plan* (DanSIS, 1988) was prepared as a Danish response to the report of the World Commission on Environment and Development (WCED, 1987), providing a perspective to the year 2000. A statistical report on nature and environment (*Tal om natur og miljø*) was launched in 1990. A further edition was published in 1994, and another is planned for 1998.

In 1992, following UNCED, the MoE decided to institute a process of broader rolling strategic planning. As part of this, the government's National Environmental Research Institute (NERI) was asked to prepare a state of the environment report. The report, *Environment and Society* (Christensen et al, 1994), reviewed media-related problems such as air, land and sea. The intention was to revise the report periodically, approximately every four years.

In 1993, the Socialist People's Party tabled a parliamentary question asking how the government intended to deal with strategic planning. In response, the Minister of Environment formally announced a process of developing strategic environmental planning, including policy and state of the environment reports for parliamentary discussion. The intention was to build on several existing elements: regular ministerial sector strategies; a range of periodic policy and state of the environment reports; and other policy-oriented papers. The parliamentary question enabled the MoE actively to promote consolidation, and move further towards an integrated approach with individual ministries retaining their responsibilities and answering to Parliament on how they are fulfilling environmental objectives in their areas of concern.

NERI's *Environment and Society* report differed from past state of the environment reviews (for example, DanMoE, 1991), which tended to focus on statistical information and contained more commentary and reflection in presenting questions for politicians to address. The Nature and Forests Agency is coordinating the development of a biodiversity action plan, while the Environmental Protection Agency is preparing a climate action plan.

Nature and Environment Policy

The scope of strategic environmental planning in Denmark was described in *The Danish Environmental Strategy* (DanMoE, 1994) (see Box 11.1). In June 1994, the MoE began to prepare a new White Paper — a nature and environment policy. The policy, *Natur og miljøpolitisk redegerelse*, was published in August 1995 (English summary, DanMoE, 1995b). It contains an overview of the main problems and environmental activities and also sets up a framework for the initiatives planned by the government in relation to policy. It sums up the state of the environment from a policy point of view and is targeted at politicians as an instrument for improving and broadening political discussions. It describes the environmental position as it exists and presents both short- and long-term visions, highlighting how to use resources in the future and setting priorities. Previously agreed objectives and actions are outlined, and measures are presented to reach new goals (for action by both MoE and other sectoral ministries). Government perspectives on future actions are also be given.

The Nature and Environment Policy, while coordinated by the MoE, is an overall government report, and represents Denmark's national sustainable development strategy. It comprises two parts: a summary discussing political questions and future initiatives (about 90 pages), followed by the main report (about 600 pages). The plan includes sections covering:

- a review of the state of environmental initiatives in Denmark;
- a discussion of long-term perspectives (a political discussion) for environmental initiatives, which considers concepts such as the global challenge, 'environmental space', cultural values, green taxes, and whether things can continue as they are;
- the main themes of future initiatives (greenhouse gases, resource consumption and health, nature, the aquatic environment, the urban environment and contaminated sites);
- individual activities and environmental pressure, covering different sectors, for example, cities, the landscape, agriculture, industry and energy supply. For each sector, the discussion includes:
  (i) historical aspects, main problems, and a list of policy actions already taken;
  (ii) already agreed environmental objectives, future policy trends and initiatives for MoE and particularly for other ministries, describing the role of MoE in relation to the responsibilities of individual ministries in revising their sector plans, and highlighting important environmental issues to be considered;
- international aspects (throughout the document, international commitments have influenced the perspectives taken, particularly those under the EU); and
- connections between environment, growth and employment.

The policy discusses issues related to trade and the environment, and cross-

Box 11.1 THE SCOPE OF STRATEGIC ENVIRONMENTAL PLANNING IN DENMARK

In The Danish Environment Strategy *(DanMoE, 1994), the scope of strategic environmental planning was set out as follows:*

It is to be implemented with a view to securing a more overall and targeted environmental policy. The planning aims at comprising all aspects of environmental problems, for example, pollution control and prevention, protection of the variety and qualities of scenery, nature and urban environments, location and land use.

Methods are to be further developed with a view to seeing environmental problems in a wider context and assessing what problems are to be addressed first, how and to what extent. The environmental policy is, to a higher degree, to be based on knowledge of, inter alia, the state of the environment, polluting discharges and other impacts on our surroundings, possibilities of technological initiatives, environmentally sound location and the economic consequences of taking action or not taking action. In accordance with the principles of prevention and precaution, environmental policy efforts will, based on an assessment of present knowledge, remain targeted at preventing potential environmental problems.

It is the intention to establish a method to continuously assess if the environmental goals may be reached through the measures agreed and if there is a need for adjustment of the environmental activities in progress. The aim is also to contribute to achieving greater integration between environment and sector policies with a view to making total environmental action more preventive and targeted. Consequently, new models, methods and decision-making routines will be developed and applied to environmental problems over a number of years as part of the strategic environmental planning.

The core concepts of strategic environmental planning are:

- *political establishment of overall goals and means against the background of continuous and targeted accumulation of knowledge; and*
- *routines for follow-up and adjustment of the environmental policy.*

The two key elements of strategic environmental planning will be:

- *an environmental progress report (state of the environment), to be drawn up at appropriate intervals, and*
- *a White Paper on the Environment, describing overall priorities, targets and specific initiatives in respect of future environmental action, and drawn up with a maximum frequency of four years.*

Moreover, the Ministry of Environment will publish topical reports on environmental matters and action, as required, in preparation for and concluding on specific policy initiatives.

The White Paper will be based on, inter alia, environmental action decisions taken at international levels, particularly under the auspices of the EU, environmental action decisions taken at the national level, actions under preparation ... and on already existing national sector action plans as well as national White Papers. For agriculture, energy and transport, sector plans have already been drawn up, which provide information on how environmental concerns are addressed in the sector concerned.

For the agricultural sector, the plan Sustainable Agriculture was drawn up in 1991. The regulatory instruments of the energy plan Energy 2000 of 1990 have recently been adjusted, and a revision is due in late 1995. The transport investment plan Transport 2005 (of 1993) is a revision of the 'Government Transport Plan of Action for Environment and Development' of 1990.

Moreover, the environment White Paper will carry forward the action plans and strategies drafted by the Ministry of the Environment in recent years, for example the plans of action on cleaner technology and on waste and recycling, a strategy for the ground water, the strategy on land use and at sea, the plan of action for the recovery of offshore raw materials, etc. Similarly, further action will be taken to promote the long-term goals for the physical and functional development of Denmark as described in the Minister for the Environment's National White Paper of 1992, Denmark on Her Way Towards the Year 2018.

In 1995, political decisions were taken on reducing CO_2 *emissions through the use of agreements and economic instruments. As a follow-up to UNCED, a White Paper will be drafted on a comprehensive Danish forestry policy as well as a strategy for the protection of biodiversity. Both are to be published in 1995. The intention is that the strategic work on key environmental issues is to be summarized in the environment White Paper in 1995.*

The Ministry of the Environment is responsible, organizationally, for drafting the overall environmental targets, in dialogue with the sector ministries as environmental concerns are to be seen in relation to and, as far as possible, as integral to other interests in society.

Within the established framework of environmental policy, sector ministries are accountable for building environmental concerns into the policies of the sector concerned. In this process, there is a dialogue between the sector ministries and the Ministry of the Environment to ensure agreement between environmental targets and efforts in the various sectors. The sector work is to be carried out in a dialogue with the sectors/actors affected, for example, business and interest organizations, counties and municipalities.

Source: DanMoE, 1994.

boundary ones such as air pollution from Germany. It also announced that the Environmental Protection Agency will develop an environmental policy focusing on products. It was written without promises of financial commitment to achieve the visions and the Ministry of Finance vetted the text. Subsequent action will be the responsibility of line/sector ministries when they revise their plans and interpret the policy (they control budgets).

Prime Motivation and Getting Going

The MoE prepared a description of the Nature and Environment Policy initiative and a timescale at the outset. A 30-page proforma or 'dummy version' outlining a possible content was prepared for discussion with other ministries. The policy is part of the development of Danish environmental policy as well as an ongoing process to follow up on Agenda 21. Its emphasis is mainly environmental as a background for holistic planning. It does not deal with social and economic issues. One of the main objectives was to provide a plan that will enable politicians to take policy decisions based on an overview of environmental problems in Denmark, and to encourage society as a whole to engage in debate on environmental policy and contribute to the strategic planning process. The government decided to engage in strategic environmental planning in 1993, and the framework of the Nature and Environment Policy was discussed by the Parliamentary Committee on Planning and Environment.

Organization and Management

The process of developing the Nature and Environment Policy was internal to government and was managed by the minister's secretariat in the MoE, but on behalf of the government as a whole. A framework for consultation and negotiation among government ministries/agencies was established. The need for the policy, the concept and its structure were discussed with all relevant ministries as a collaborative venture. Most ministries preferred the MoE to take the lead in writing sector chapters and they would then review and comment on them. Some ministries made initial contributions. Draft sections were discussed and negotiated in numerous meetings.

As mentioned earlier in this book (p40), a working group made up of personnel from the Ministry of Environment and three of its agencies (namely the Environmental Protection Agency, the Nature and Forests Agency and the Energy Agency) and experts from environmental and geological research institutes, carried out the main body of the drafting work. Other working groups were also established within each of the individual agencies. Individuals drafted particular elements of the policy plan. Draft materials were reviewed within the working groups as a whole, and were regularly examined by a higher-level meeting of heads of MoE divisions acting as a network steering committee or management core group (meeting every two weeks). Difficult issues and conflicts were referred for resolving to a committee of directors-general of agencies.

In the initial stages, dialogue involved the various agencies of the different ministries. For the final stages, a seven person drafting/editorial team, comprising the best writers from the different agencies, was established in the MoE. This team engaged in intensive dialogue with the Minister of Environment's secretariat and, after the discussions and negotiations, there was agreement across government on the text. After publication, the Nature and Environment Policy was discussed by the Parliamentary Committee on Planning and Environment, and it was intended to discuss the proposals with NGOs, other organizations and industry, particularly in terms of future initiatives.

Key 'Assistance' Factors and Problems

Two important factors helped the development of the new policy. First, the preparation of the 'activities and loads' sections generated enthusiasm among ministries because the contributors were able to address broader issues than their responsibilities normally allowed. And second, there is a tradition in Danish policy-making to take the environment seriously, and a willingness throughout government and in Parliament to face and deal with environmental issues. But a major problem was dealing with the question of how to sustain current lifestyles in Denmark.

Local Agenda 21s

In November 1994, the MoE, together with the heads of the Association of Municipalities and the Association of Counties, wrote to the leaders of all counties and municipalities urging that Local Agenda 21s be started. Some local municipalities and authorities are now developing such local strategies. But the Nature and Environment Policy is not linked to these. However, it was sent to all local authorities to influence their Agenda 21 processes. The MoE is also encouraging them and is providing advice. A working group has been established comprising representatives from the Association of Municipalities. In collaboration with the Association of Municipalities and the Association of Counties, the MoE has produced a guide on preparing Local Agenda 21s, *Local Agenda 21: An Introduction to Counties and Municipalities* (DanMoE, 1995a). This provides concrete examples and explains how to organize the process. It is targeted at local politicians and civil servants. The three organizations intend to organize seminars and conferences to discuss opportunities and to exchange experiences.

Some of the MoE authors of the guide also contributed to writing the urban section and political introduction to the draft Nature and Environment Policy, providing some linkage between the two processes. Policy-making in Denmark is very decentralized, with considerable dialogue between civil servants and politicians at national, county and municipality levels.

Parliamentary Process

The Nature and Environment Policy was first presented to Cabinet for formal approval on behalf of the government. The Minister of Environment and Energy submitted it to the Parliamentary Committee on Planning and Environment, and it is hoped that a wider debate will ensue. The policy was to be distributed to NGOs and society groups to stimulate responses, and the aim was that the document would be used by sector ministries in developing their further policies, programmes and plans.

Greening Lifestyles and Mainstreams in Denmark

At the level of the civil service, the Nature and Environment Policy process has helped to broaden perspectives and to increase general environmental awareness outside routine areas of responsibility. For example, it has helped the Ministry of Defence and Danish Railways to produce their own green plans.

Chapter 12 FRANCE

The two main strategy processes in France are discussed in this chapter: the National Plan for the Environment (*Plan national pour l'Environnement* — PNE), published in 1990; and the work of the French Commission for Sustainable Development (*Commission française du Développement durable*), decreed in 1992 and established in 1993. Other relevant processes include an internal exercise, currently underway within the Ministry of Environment, to develop a scenario for sustainable development in France for the years 2010 to 2030. In addition, some 20 commercial businesses have prepared or are developing voluntary environmental plans (for example Renault and Elf Acquitaine). Successive five-year development plans — 11 have been prepared since the Second World War — maintain a balance between France's various regions. A National Plan for Sustainable Forest Management was adopted by the government in April 1994.

Plan National pour l'Environnement (PNE)

Prime Motivation and Getting Going

Following the re-election of President Mitterrand in 1988, Michel Rocard became Prime Minister and a more open path of government was pursued. As part of this process, the PNE was initiated in early 1989 by the Prime Minister and the then Minister of Environment, Brice Lalonde, partly as a demonstration of their personal commitment to the environment. The terms of reference for the PNE were effectively set out in a letter from the Prime Minister to Brice Lalonde dated 2 November 1989, which affirmed support for ideas for such a plan presented by M Lalonde to the Prime Minister in October 1989. The ideas laid out the need for structural change in the Ministry of the Environment (MoE).

The PNE had a ten-year perspective and was to be prepared within about six months. One of its main aims was to strengthen the structure of the MoE and to give weight to environmental policy. This was to be achieved through increasing the staff complement and budget, providing a more independent structure with regional offices and regional direction, and increasing the ministry's profile and influence.

Prior to the PNE, many ministerial services for the environment were spread over several sector departments. The MoE was more concerned with coordination than with management and the same model of strong sectoral departments oriented towards investments was duplicated at the district (*département*) and regional levels of government. The regions were more policy oriented with a 'mission to coordinate investments at lower levels and to monitor economic and social well-being' (French MoE, 1990c). The PNE aimed to introduce reorganization, particularly for work at the regional level. The idea of establishing both regional- and district-level offices of the MoE was suggested and discussed, although, to date, only the regional offices have been created. There was considerable opposition to these changes in sector ministries and departments, for they represented a loss of power.

The French MoE (1990c) has described the PNE as a 'charter for public and private, corporate and individual actions, with accompanying financial obligations', and has stated that the object of the PNE is that

> *the principle of long-term social viability becomes the fundamental reference point for decision-making for the French society as a whole, and in particular for industry and government. Success requires a commitment to make up for lost time as well as to change behaviour.*

Focus

The overall objective of the PNE was 'the integration of the environment within all aspects of the society: through reform in public administration, the economy, taxation, industrial and social relations, and cultural behaviour' (French MoE, 1990c). Its principal focus was 'environmental', dealing with such issues as waste and pollution, developing 'horizontal' approaches such as eco-taxes, and focusing on the responsibilities of other ministries concerned with the environment.

The PNE presented a diagnosis of the state of the environment in 1990, the probable situation in the year 2000, the positive and negative attributes of environmental policy in France up to 1990, and the importance attached to environmental questions. In addition, the PNE announced eight principles for action to assist coming to terms with the environmental crisis in France and elsewhere (see Box 12.1). The public version of the PNE (French MoE, 1990b) contained a chapter that specifically addressed 'greening' policies and the economy.

The NPE employed 'environmental quality' as a catalyst for innovation and economic competition. It first aimed to catch up with past problems and address the questions of risks, set out to establish a policy of preventing environmental damage in the future as well as restoring degraded natural areas, provided for overall reduction in negative environmental forces, and sought to attenuate the ecological inequalities in a societal sense. This approach contrasts with orientations for sustainable growth as expressed in some other national policies (for example, the Netherlands and Canada).

Domestic sectoral objectives were organized under: air, water, marine and coastal environment, wastes, nuclear and civil security, control of chemical products, noise, landscape and nature conservation, and urban ecology.

The PNE aimed to promote partnership in action between different levels of government or public and private interests. It envisaged up to 200 environmental action plans to address and resolve urban environmental problems, made at communal and intracommunal levels (see section below on provincial, local and convention-related strategies). A summary of the PNE is presented in Box 12.2.

Box 12.1 PRINCIPLES FOR ACTION IN THE PNE

- *environmental quality is a positive factor for economic competitiveness;*
- *costs will be reduced through innovation and prevention;*
- *rigorous regulatory enforcement by a modernized state administration;*
- *delegation of responsibilities to local government as well as others through partnership, as a means for the environment to become everyone's affair;*
- *improvement of knowledge and skills as the basis for decisions which will be more rational and cost-effective;*
- *testing of new ways for democratic participation in public decision-making at both the national and local levels;*
- *affirmation of social justice and solidarity through the use of the 'polluter pays' principle and the reduction of ecological inequalities; and*
- *embracing the international concern for the sustainable use of the earth's natural resources.*

Source: French MoE, 1990c.

The NGO and public response to the PNE was generally favourable, but there was a predictable feeling that the plan could have been more ambitious.

Organization and Management

The principal writers of the PNE were Lucien Chabason (a director in Brice Lalonde's 'cabinet') and Jacques Theys. These two coordinated a 12-person PNE preparation team (mostly of Ministry of Environment staff). Formal consultations and numerous meetings were held with and contributions received from all relevant government ministries, agencies, regional delegations on architecture and on the environment, and the Federation of Nature Parks. The plan evolved as an interdepartmentally negotiated text.

Informal meetings and discussions were held with industries, businesses and NGO associations. Written contributions were requested and received from NGOs on their ideas for inclusion in the PNE about their role in the decision-making process. Most of these ideas were taken into account. There were also some informal discussions with representatives of various cities and with a range of politicians and trade union leaders.

About midway through the development of the PNE, a public meeting was organized for some 300 invited people. It was attended by the President and Prime Minister, and by all ministers who were asked by the Prime Minister to present their ministerial responsibilities, policies and actions concerning the environment. This was the first occasion in France that such a meeting on the environment had been held. Open meetings were later held in seven of France's 22 regions at which the draft PNE was presented for discussion. Apart from these consultations, the process was entirely internal within the Ministry of Environment.

There was consensus across government on most of the general issues but not on each proposition. The PNE was presented to Parliament in September 1990 to facilitate an 'orientation debate'.

Box 12.2 SUMMARY OF THE PNE

AMBITIONS	ACTION
■ Stop production and use of CFC within ten years;	■ Yes (ii)
■ Stabilize CO_2 emissions in medium-term horizon, 2000–2005;	■ Yes (i)
■ Reduce atmospheric pollution by 20 to 30 per cent;	■ New law in preparation
■ Increase from one-third to two-thirds the level of domestic waste water treatment;	■ New water law (1992); and water agencies' sixth action programme
■ Treat and reduce pollution coming from agriculture (particularly from nitrates);	■ Implementation of European directive on nitrates
■ Limit to minimum possible the disposal of wastes in landfills, and raise to 50 per cent the level of reuse and recycling of primary materials (as compared with one-third at present);	■ New law (1993)
■ Isolate or renovate 200,000 dwelling units still subjected to intolerable noise levels;	■ New law (noise) (1993)
■ Achieve significant reduction in vulnerability to technological or natural risks;	■ New law (natural risk prevention) (1994)
■ Conduct a determined and rigorous policy to protect nature and landscapes over the whole nation, giving particular attention to certain fragile areas (coastline, wetlands, woodlands next to metropolitan areas, high and medium mountains, overseas dominions and territories); and	■ New law (landscape) (1993); new coastal conservation organization
■ Use all means (financial, institutional and scientific) to resolve planetary scale environmental problems (especially western and eastern Europe, Mediterranean basin and Pacific).	■ Funds to GEF
THE MEANS	
■ Constitute an enlarged and autonomous MoE, with its own services and policy sectors regarding water, the prevention of pollution and risks, and the protection of nature and the landscape;	■ Yes (i)
■ Create regional offices of the MoE (Direction régionale de l'Environnement – DIREN)	■ Yes (ii)
■ Better use of the territorial administrations;	
■ Create a corps of agents deputized to render police service for all aspects of legislation concerning the natural environment;	■ No

■ Create a French Institute for the Environment (information only); and a National Institute on the Environment, Risks and Industrial Security;	■ Yes (ii)
■ Organize technical and university-level training to embrace skills and knowledge related to the environment, resulting in new professional qualifications;	■ No
■ Develop within the public administration a corps of environment inspectors;	■ No
■ Establish a specialized programme for training engineers and senior-level administrative staff with regard to environmental matters;	■ No
■ Increase public and private investment in applied research for the environment by 100 per cent;	■ Yes (i): About 30 per cent achieved
■ Create a fund for stimulating partnership in environmental improvement schemes, financed by taxes on gravel extraction, signboard publicity and motorway traffic ('a centime per km travelled');	■ No
■ Extend or reinforce the 'polluter pays' principle to wastes, noise, nitrates and the pollution responsible for the greenhouse effect; and	■ Partially (air, water and wastes)
■ A threefold increase of the MoE budget in five years.	■ Doubled
THE METHODS	
■ Formulate an environmental code of legislation;	■ Yes (ii)
■ Increase the severity of sanctions for 'ecological delinquency';	■ Yes (i)
■ Clearly separate, to the extent possible, the administrative roles in project development and control;	■ No
■ Give Parliament an increased role in formulating major decisions about the environment (including participation in the formulation of European regulations and international conventions), and extend the operations of the Parliamentary Office for the Evaluation of Technological Choices;	■ Under discussion
■ Extend environmental impact assessment procedure to cover major legislation and national investment programmes;	■ In project (connected to European Directive)
■ Improve environmental impact assessment procedures and public debate;	■ Yes: Commission du Débat public; plus Circulaire BIANCO (internal operational guidelines on administration of public debate for infrastructure investments)

■ Create regional-level committees for technological evaluation, playing a role in mediation and in financing alternative expertise;	■ No
■ Confer responsibility for landscape policy to the regions;	■ Some general responsibilities transferred
■ Launch, within five years, 200 communal and intra-communal environmental action plans;	■ Yes (ii)
■ Develop partnership between national government and local authorities with regard to new financial resources;	■ No
■ Render obligatory 'ecological audits' within large enterprises, having the same status as 'social audits';	■ Not compulsory
■ Confer to ANFOR (national agency for norms and certification) the creation of an 'eco-product' label, compatible with an eventual European label;	■ Yes
■ Enlarge remit of health and work safety committees to include problems linked to both risks and the environment;	■ No
■ Progressively establish an 'ecological fiscality' by adding an environmental dimension to existing taxation; and	■ No
■ Develop a policy for the use of criteria concerning 'clean' products and technology, or resource savings, in the adjudication of tenders for public-sector work.	■ No
THE NEED TO MASTER THE CONSEQUENCES OF ECONOMIC GROWTH	
■ Harmonize energy taxes at higher end of spectrum;	■ Increased taxes on gas and oil
■ Create a fund to promote renewable energy sources;	■ No
■ Develop combined rail/road transport;	■ Yes (but at slow pace)
■ Private partnerships for recycling plastic wastes	■ Yes (eco-emballage system)
■ Limit traffic circulation in town centres of large urban agglomerations;	■ Yes (by-laws in preparation in some towns)
■ Establish mechanisms for financial solidarity of benefit to communes or farmers that assure the management of the natural heritage;	■ Local plans for sustainable agricultural development
■ Develop incentives for energy savings and material recycling;	■ No
■ Better integrate the environment in policy concerning run-down neighbourhoods; and	■ No
■ Create a National Agency for Environment and Energy Saving	■ Yes

Source: French MoE, 1990c.
Notes: ADEME = Agence de l'Environnement et de la maîtrise de l'Énergie. (i) = action in progress. (ii) = action fully implemented.

The PNE represents a set of proposals; it was not law and was not binding. However, successive governments have implemented elements of the plan. To date, about two-thirds of the measures proposed in the PNE have been implemented, for example, through a range of new laws concerning water, waste, noise and landscape.

Key 'Assistance' Factors, Problems and Issues

The development of the PNE was greatly assisted by the keen interest of the Prime Minister, the fact the M Lalonde was an ecologist himself, and the strong public interest concerning the environment: 'The public opinion today is far more demanding with regard to environmental quality than it was twenty years ago. The public also recognizes that the potential impact of environmental quality on the economy has changed' (French MoE, 1990c).

There were, however, a number of problems in developing and implementing the PNE:

- The other ministries were mainly sceptical because the MoE was a relatively 'small' ministry with negligible influence, power, credibility or perceived legitimacy. They found it difficult to understand or accept that the environment was either a serious or a legitimate issue.
- The weak position of the MoE in government at the time made it difficult to implement the plan. The tradition of 'planning' had also become less fashionable in France.
- Implementation of the PNE as a long-term plan was hampered and interrupted by lack of continuity between governments.
- While two-thirds of the PNE has now been implemented, some key ideas have proved difficult to get adopted; for example, the MoE proposed to assume responsibility for nuclear security, but this is still mostly under the Ministry of Industry; also, proposed environmental taxes on motorway use and on large shopping centres — to flow to a special fund — have not been taken up.

Key issues the PNE sought to address included: fresh water pollution, particularly from agriculture (for example, nitrates); urban air pollution; waste management; and ecological 'infrastructures'.

Links with National Planning and Decision-Making

The PNE's objectives are reflected in the eleventh French five-year development plan. To diversify the French regional planning system, the government has decided, through a new law on *aménagement du territoire*, to establish territorial development directives. Drafted with the major local authorities, they will be approved by the state, and will aim to enable it to ensure respect, at the proper scale, of the necessary balance between development prospects and the protection of natural areas, sites and landscapes (French MoE, 1995).

Provincial, Local and Convention-Related Strategies

About 25 towns or cities and several *départements* (regions) have signed contracts with the MoE to develop 'global' environmental plans (identifying priorities) under environmental charters (*chartes pour l'environnement*). The MoE has produced a guide to assist towns in this process (French MoE, 1994) and also provides some funding and assistance. The aim is to undertake a comprehensive environmental protection approach: protection of natural resources, of fauna, improving landscapes, urban and rural heritage, waste elimination and prevention of natural risks. Under the contract,

> *the state undertakes to contribute its technical and financial assistance to the local authorities who so desire. In turn, these agree to comply with the principles and goals of the Environmental Charter and to initiate a three-staged approach: environmental diagnosis, defining concerted priority goals, and drafting a programme of multi-annual actions.*
>
> (French MoE, 1995)

The MoE produced a national plan for climate change in December 1994 and a biodiversity plan is in preparation. Neither of these is formally related to the PNE.

Dependent Territories

France has three *départements d'outre-mer* (overseas departments) — Réunion, Martinique and Guadeloupe — in each of which a new regional office of the national MoE has been established following the PNE.

It is also responsible for three dependent territories — New Caledonia, Tahiti, and the Wallis and Futuna Islands. Each has its own MoE but none has yet prepared a sustainable development plan, although an environment plan is being developed in Tahiti. Environmental profiles have recently been completed for each territory (IFEN, 1995a, 1995b and 1995c).

Ecological Footprints and Transboundary Issues

The PNE did not focus on ecological footprints and transboundary issues, although it does contain a chapter on European cooperation in which the very significant influence of European policy on national policy-making is discussed. There is also considerable discussion of climate change issues, North–South relations and of France's commitment to the Mediterranean Action Plan (France provides 40 per cent of the budget).

Press Coverage

The PNE was generally well received by the press. Following publication of the official PNE document, a public version was prepared which contains extracts of press commentaries (French MoE, 1990b). These tended to concentrate on the proposed structural changes for the MoE.

Follow-Up

In early 1994, the MoE began an 18-month study to explore scenarios for sustainable development in France in the years 2010 to 2030. The resulting report will be internal to the MoE and not a government plan. It will cover macro-economic, sectoral and international aspects and will focus on two separate scenarios: (a) one resulting if current trends persist; and (b) another based on following a sustainable development path. The MoE is trying to assess the possibilities of achieving transition from (a) to (b) by investigating the 'margins of manœuvrability'.

The work is being carried out by a core group of 15 MoE staff and experts working part time on the task. Two meetings are held each month. Specialist subgroups (five to seven people) have been set up to deal with sector aspects and individual topics, and these also include experts from universities and government (the latter sitting as experts themselves). No external consultations have been held. The report was to be published and presented to the French Commission for Sustainable Development in early 1996.

In October 1995, it was announced that a national strategy for sustainable development will be developed during 1996 under the guidance of the Prime Minister to ensure that it is well integrated into public long-term plans. The strategy exercise will be carried out by the MoE in conjunction with other concerned ministries. Its aim will be to define appropriate objectives so as to orient the actions of actors and to define indicators to measure progress. The conclusions will be presented within the framework of a national meeting on sustainable development, scheduled to take place at the end of a national debate during 1996, and socioeconomic actors, associations and citizens will be allowed to express their opinions. The French Commission for Sustainable Development (see below) will play a full part in this process.

French Commission for Sustainable Development

Initiation

The idea of the French Commission for Sustainable Development (FCSD) was first decreed in 1992 with a proposed composition of 56 members drawn from across society (from the civil service, industry and NGOs), with a role to advise the government on UNCED follow-up. Following the change of government in 1993, the composition was considered to be too wide and a new modified structure was introduced in January 1994 under Decree No 94.65. This formally set up an FCSD comprising a chairperson and only 15 members representing scientists, experts, business leaders and representatives of associations (there are no government members). Members are nominated by the Minister for the Environment and the MoE has appointed an officer responsible for liaison with the FCSD. It was formally initiated on 20 April 1994 and works very closely with all government ministries, which it requests to undertake studies within their areas of competence. Financial support is provided from the MoE and from the Prime Minister's Office.

The first president of the FCSD was the industrialist Bernard Esambert, appointed in September 1994. He was succeeded in January 1995 by Jean Pierre Souviron, also an industrialist.

Focus

The FCSD advises the Prime Minister on implementing Agenda 21. In France's 1995 report to the UN Commission for Sustainable Development, the role of the French CSD is described as:

> *To help the government draft its strategy [for sustainable development], contribute advice and the required instruments to translate into facts the major principles of Agenda 21.*
>
> *The action of the state, by tradition, is normative or regulatory. In order to mobilize civil society on sustainable development, the state must go further, launch and promote a new approach based on dialogue, contracts, partnerships. In searching for a higher quality growth, which saves resources and space, the French Commission has a duty to be imaginative: it must offer society new modes of operation, and the government new terms and conditions for intervening.*
>
> (French MoE, 1995)

The FCSD is able to select its own work areas, make agenda proposals and seek solutions. It is able to access and benefit from other studies carried out by the Planning Commission and to integrate with these. But it has yet to succeed significantly in raising awareness about and/or acceptance of the concept of sustainable development within the planning commission

M Esambert (1995) defined ten priority themes for the FSCD to focus on during 1994/5. These were considered to be too many by M Souviron and the Prime Minister. In a letter from the Prime Minister in January 1995, the FCSD was requested to focus on two key issues:

- to define indicators for sustainable cities and towns in collaboration (now being undertaken with the French Institute of the Environment — IFEN); and
- to consider urban ecological issues and provide advice on developing urban sustainable development strategies.

In 1995/6 the FCSD also reflected on various other issues:

- sustainable development and demography;
- democracy through cultural development;
- economic and financial policies in the area of environment and sustainable development;
- indicators of sustainable development (in general);
- sustainable development and job creation;
- consumption models;
- possible establishment of a CSD in French Guyana (if wanted locally); and
- transfers and change North–South–East.

Consideration was also being given to pursuing two other important areas of work:

- The development of criteria for aid — France's aid contribution is near to 0.7 per cent of GNP, but is not used efficiently; and
- The mobilizing and coordination of research ideas on sustainable development, setting clear objectives for public research on sustainable development.

The FCSD will play a full role in the preparation of the national sustainable development strategy during 1996.

European Conference of the National Commissions on Sustainable Development

In January 1995, in Courchevel, the FCSD hosted the first conference of National Commissions for Sustainable Development from European and North American countries (FCSD, 1995).

Chapter 13 LATVIA

Izabella Koziell

There are two main national processes in Latvia: the *National Environmental Policy Plan* for Latvia; and the National Environmental Action Programme. The Environmental Action Programme for Central and Eastern Europe has also been influential.

Initiation and Time Perspectives

The Ministry of Environmental Protection and Regional Development (MEPRD) was responsible for coordinating the preparation of the *National Environmental Policy Plan* for Latvia (NEPPL) — a long-term strategy with an approximate time span of 20 to 30 years, as well as the National Environmental Action Programme (NEAP) — which will incorporate short-term actions of one to five years that are necessary to implement the NEPPL.

The NEPPL process was initiated in December 1993 and the final version of the plan was produced towards the end of January 1995 (then revised and agreed with sectoral ministries in March 1995, and accepted by Cabinet on 25 April 1995) (MEPRD, 1995). The NEAP was started in November 1994 and was due to be ready by October 1995.

While the two strategies were developed concurrently, each had its own clearly set objectives, and they did not overlap either in scope or in content. These strategies were the first to be developed in Latvia since the founding of the MEPRD in June 1993. The State Committee for Nature Protection — established in 1988 and becoming the Environment Protection Committee in 1990 — had played no significant role in environmental strategy development.

The Environmental Protection Department's director and core group leader has estimated that the core group members expended a total of two man months directly on the development of the NEPPL, while the process involved an overall total of 450 person days in 1994 and 60–100 more in 1995.

Prime Motivation

One of the main motivating factors behind the development of the NEPPL and the NEAP was the need to produce a base for sustainable development in Latvia by integrating environment into all key sectors. Since 1990, the environmental protection system had been developing rapidly. However, no comprehensive plan outlining the goals, priorities, problems and mechanisms of environmental policy had been adopted by the Cabinet of ministers or Parliament. It was recognized that there was a need to develop a concise and clearly presented environmental policy (NEPPL) and National Environmental Action Programme (NEAP).

It was also recognized that there was a need to raise environmental awareness within the key sectoral ministries to help instil commitment to environmental protection. The NEPPL process aimed to heighten awareness to help lead to the adjustment of sectoral policies so that they incorporate environmental protection concerns. The need for clearly defined environmental policy has been demonstrated by past experiences of uncoordinated management, and a lack of cooperation and communication between ministries, resulting in the *ad hoc* implementation of actions based on 'highly scientific' and generally irrelevant priorities.

Other motivations included the need to prove to the international community that Latvia is a serious partner committed to sustainable development and hence to assisting in the struggle to attract investments, soft loans and grants. It was felt that a clearly presented policy document would help international financing agencies and donors assess the need of assistance and ensure that assistance was not misdirected.

It was considered that the NEPPL document would also help private companies and investors understand the implications of the new environmental requirements on business and industry, and would give guidance on the enforcement of new standards and legislation. This is particularly important for Latvia, which is undergoing an economic 'transition' with all its associated changes (such as privatization), and where general economic decline has resulted in short-term interests taking precedence over long-term ones, leading to over exploitation of natural resources and a lack of environmental rigour. Furthermore, the NEPPL was seen as a means of assisting Latvia achieve compliance with international commitments and national legislation. It was also aimed at explaining what needs to be done by the general public to improve the state of the environment.

The NEAP was needed specifically to define what actions were necessary to implement the NEPPL, to plan the rational use of a limited budget, and to outline the budgetary requirements. It functions as a background for negotiations with private companies on investments in the environment.

Focus

The NEPPL and NEAP do not have one single focus. Given the complex nature of environmental problems, the challenge was to develop a fully integrated and interdisciplinary approach to environmental protection. The main focus of the NEPPL and NEAP was to provide a base for integrated planning and implementation for sustainable development. The NEPPL document outlines four environmental policy goals, ten priority problems and ten environmental policy principles. Box 13.1 lists NEPPL goals and priority problems.

Organization and Management

The MEPRD has been fully responsible for the strategy processes. The NEPPL project

Box 13.1 NEPPL GOALS AND PRIORITY PROBLEMS

Environmental Policy Goals

1. *Substantial improvement of environmental quality in the areas where it generates significant risks, at the same time sustaining environmental quality in the rest of the territory.*
2. *Maintaining present levels of biodiversity and preservation of traditional landscape patterns.*
3. *Sustainable use of natural resources.*
4. *Safeguarding integration of environmental policy into all sectors of development, and raising environmental awareness of the decision makers and general public, hence providing a basis for sustainable development.*

Priority Environmental Problems

- *Transboundary environmental problems (air, water)*
- *Eutrophication of watercourses and degradation of aquatic ecosystems*
- *Risk caused by economic activity*
- *Problems caused by generation and accumulation of waste*
- *Problems caused by transport*
- *Problems caused by agriculture*
- *Depletion of biodiversity*
- *Degradation of landscapes*
- *Depletion of natural resources (renewable and non-renewable)*
- *Low quality drinking water*

Source: MEPRD, 1995.

was initiated in 1994 by the Minister for the Environment and responsibility for completing the work was delegated to the Environmental Protection Department (EPD). A core group of four persons (all from the EPD) was created and led by the EPD's director. Eleven subgroup members were selected from the EPD, the Division of State Cadastres and Natural Resources and from a city planning company. The project group (60–70 persons) included representatives from the different sectoral ministries and institutions. Participation by key target groups and NGOs was limited (only those groups existing at the time were involved, for example the Energy Agency and the Latvian Fund for Nature). There is still a general lack of target groups and few NGOs existing in Latvia, a legacy of Soviet rule.

The NEPPL was developed with assistance from Dutch and Swedish environmental specialists, with advice provided on process rather than content. The method used was proposed by the Dutch Ministry of Housing, Spatial Planning and the Environment, since there was no previous experience of interactive methodology existing in Latvia.

The NEPPL process began with an initial brainstorming session within the MEPRD, which resulted in a list of key environmental issues and problems. This was followed by a series of 'interactive' seminars involving the core and project groups, where ideas were noted on flip charts and later used to compile a draft document. The seminars brought together experts from a wide range of sectors and aimed to gather their expertise and integrate all opinions and views into one document.

The NEAP is the implementation action programme of the NEPPL and no real implementation management structures are yet in place. The need to improve communication between ministries and between target groups has been identified. The process of implementation remains complicated within the rapidly changing administrative and legislative environment in Latvia.

Since the NEPPL sets out the policy goals over a period of 15 to 20 years, it is to be revised in accordance with the election regime (ie every three years starting in the autumn of 1995). It is envisaged that the NEAP will be revised every year, or every second year, to keep it in tune with new priorities and problems.

Terms of Reference

Originally, there were no terms of reference set for the work — there is no legal requirement in Latvia to set these. The process and concepts for both the NEPPL and NEAP were discussed at length and agreed within the MEPRD, and the minister set a short regulation for these. The core group was given full authority to develop the approach and management mechanisms. At the request of the Dutch, who funded part of the seminar costs, a project plan was prepared by the core group in English.

Participation

The NEPPL process involved members of the Environmental Protection Department, representatives from the Division of State Cadastres and Natural Resources, the different sectoral ministries (including economics, finance, agriculture and the state forest service) and environmental institutions (the Institute of Biology, the Latvian University Ecological Monitoring Centre, and the Sea Monitoring Centre), and those target groups that existed at the time (for example, the Energy Association). Two private consulting companies and Riga City Environment Department were invited. Two of the main NGOs, the Latvian Fund for Nature and the Environmental Protection Club, also participated. The same group of persons was involved in the NEAP process, with the Department of Projects of MEPRD — responsible for project management and implementation — playing a central role. NGOs were invited to comment on the drafts and participated in seminars. Comments on drafts of the NEAP were sought from the few active NGOs in Latvia, and they were invited to participate in seminars.

The initial NEPPL brainstorming session resulted in the formulation of draft policy principles, a summary list of priority problems and policy actions. The seminars that followed were both consultative and participative in that those who attended were initially asked to comment on the draft and incorporate any issues that had been omitted. The second seminar focused on the selection and development of a suitable set of environmental policy instruments and the third involved discussion of the final draft. Some participants from the project group have also been involved in drafting of the document.

The NEPPL was rather weak on public participation and it proved difficult to incorporate public views. Organizations representing NEPPL target groups or sectors, such as farmers and industries, were few and poorly developed and it was difficult to access opinions. The legacy of socialist psychology and thinking appears to have had a detrimental effect on people's desire and interest to participate; generally, local people are not interested in 'doing', but prefer to wait for the government to solve all their problems. After 50 years of Soviet domination, it will take time for the socialist mentality to disappear and for new democratic attitudes to develop.

The NEPPL has failed to stimulate communication between the environmental authorities, the mass media and the public. The mass media are only just beginning to develop their techniques and learn to target key issues of interest to the public. Although there was some publicity surrounding the NEPPL, on the whole the public received very little information about the policy and action programme. The MEPRD's Public Relations Department has, however, developed a communications strategy as a supplement to the NEPPL.

Generally, the consumers' main interest at present is to 'get rich quick' at the expense of the environment. Also, for many businesses, it is not currently economically viable to implement 'sound environmental management practices' and they are not interested in supporting any strategy. Most are waiting for the state to provide such assistance, which at the moment it cannot do. The prevailing psychology is such that, although the public are free to participate, they do not want to and some are not even aware of their new freedom to do so. There was also too little funding to enable full participation.

Key 'Assistance' Factors and Problems

No specific factors assisted the development of the NEPPL. There was no existing strategy and no real public pressure. On the contrary, in the light of recent political and economic changes, some members of the public have a very negative view of environmental protection, especially where it begins to affect their livelihood capabilities. Indirectly, pressure from international partners (for example, the Dutch donors' stipulation about timing and the World Bank NEAP requirement for receipt of future investments) was positive and helped push the whole process forwards faster. There was considerable moral support from some political leaders and, although not all staff in the MEPRD were interested at first, the great enthusiasm displayed by the core group helped generate wider interest in the whole process.

One of the major problems was lack of time. All those engaged in producing the NEPPL document had other responsibilities within their departments, which meant that the core group was unable to work full-time on it. Furthermore, the members of the core group, all of whom were from scientific backgrounds, had no previous experience of strategy development, or of participatory and interactive approaches.

The lessons learnt during the NEPPL process are being applied to the NEAP. And financial resources are being sought from national organizations in an attempt to resolve all the important and key issues.

Conflict resolution

No real conflicts emerged during the NEPPL process. After so many years of limited freedom of speech, participants respected the opportunity of freedom to express their own ideas. To avoid any misunderstandings, the 'rules of the game' were agreed upon in the initial stage. It is likely that more conflicts will occur during the NEAP process, especially since budgetary allocations are likely to be involved.

Links to National and Local Planning, and Decision-Making

National- and local-level planning systems are still relatively weak in Latvia, as are the linkages between national- and local-level government departments. As a result, the NEPPL process was developed independently of any national or local planning systems. Furthermore, due to a serious lack of funds at the regional level — barely enough to cover salaries — the regional departments give priority to their environmental inspection duties and planning remains of secondary importance.

The Environmental Action Programme for Central and Eastern Europe (see Chapter 6) is of some relevance to Latvia, though its greatest benefit to the MEPRD has been in its utility as a 'handbook'. While the NEPPL was able to draw from some aspects of it, other parts of it are less relevant to Latvia than to other CEE countries (for example, its emphasis on human health and air pollution).

There are no real local-level strategies in existence yet, but members of the core group have on occasions been invited to regional planning exercises to assist in facilitation. Ventspils City recently developed an environmental protection plan with emphasis on strict environmental licensing.

Convention-Related Strategies, Ecological Footprints and Transboundary Issues

Latvia has ratified seven important conventions. Generally, it is easier to authorize the 'low cost' ones, for ratifying conventions has serious financial, administrative and institutional implications and places a great burden on existing staff whose time is already limited. The Nature Protection Division has been made responsible for developing the national Biodiversity Action Plan (BAP) and great effort is being made to ensure that this is relevant to Latvia and that it ties in with the NEPPL and NEAP process. The person responsible for the BAP has been a member of the NEPPL core group and this allows for easy flow of information. The Nature Protection Division has been assisted by WWF Sweden.

Great effort has been made to make the NEPPL and NEAP pragmatic and 'implementable'. Ecological footprints have not been looked at *per se*.

Latvia has not yet signed all the relevant conventions relating to transboundary pollution, but there is minimal transboundary pollution emerging from Latvia. It is waiting for neighbouring countries, which produce a considerable amount of pollution that affects Latvia, to sign first. There is no specific agreement concerning Latvian-bound pollution between Lithuania and Belarus.

Parliamentary Process, Wider Debate and Press Coverage

There has been considerable debate in the Cabinet, but the NEPPL has not yet been submitted to Parliament. The MEPRD wants it to become well established before taking it there. There is also concern that since there is no specific 'green' interest in Parliament, it is still relatively new and inexperienced, it could undermine the NEPPL. At present, the MEPRD feels that government approval is more vital and more valuable.

The NEPPL process has created more environmental awareness among the participants of the seminars. Since the plan has not yet been implemented, it is still too soon to judge its external impact. The process has, however, helped technical specialists to think in more multidisciplinary terms.

The NEPPL and NEAP processes have received very good coverage on television and in the newspapers.

Chapter 14 THE NETHERLANDS

The directorate for strategic planning in the Ministry of Housing, Spatial Planning and the Environment (VROM) has coordinated and prepared a series of National Environmental Policy Plans — NEPP (1989); NEPP+ (1990); NEPP2 (1993). During the NEPP's preparation, the government's National Institute of Public Health and Environmental Protection produced an influential report, *Concern for Tomorrow* (RIVM, 1989), which assessed the state of the environment and connected this with the ultimate goals needed for a sustainable future in the Netherlands.

In 1992, Milieudefensie independently published the *Action Plan: Sustainable Netherlands* (English translation, 1993). Also in 1992, the Ministry of Agriculture, Nature Management and Fisheries developed a plan for the ecological structure of the Netherlands that links important ecological areas with each other, subsequently approved by Parliament in 1995 (LNV/VROM, 1995).

In the Netherlands, physical development plans are prepared regularly every ten years. The last one was completed prior to the NEPP. Efforts are now being made to see whether it is possible further to integrate physical planning with the process of developing NEPP3 in 1996/7.

Informal discussions about the environment with business and industry have been undertaken by Natuur en Milieu (the Nature And Environment Foundation). The Institute for Environment and Systems Analysis (IMSAR) has also organized round tables, while between December 1994 and March 1995 Milieudefensie organized informal round tables for key actors from business, industry, politics, science, the economy and trade unions.

National Environmental Policy Plans

Initiation and Time Perspectives

The first *National Environmental Policy Plan* (NEPP) was initiated in 1987 and published in 1989. According to the Ministry of Housing, Spatial Planning and the Environment (VROM, 1993b),

> *It pioneered a detailed, target-based approach to the management of environmental problems facing the country. Its ambitious objective was to reverse unsustainable development within one generation.*

As stated earlier (p21), this was a shift away from the effect-oriented and rather compartmentalized approach that characterized 1970s and 1980s to the more integrated approach of themes like:

- climate change;
- acidification;
- eutrophication;
- diffusion (in other words the uncontrolled spread of chemicals, often by dumping of toxic waste, into the environment);
- waste disposal;
- disturbance (noise and odour);
- groundwater depletion (which has a detrimental effect on water supplies and natural habitats); and
- squandering of natural resources.

As mentioned earlier, VROM decided to implement this change of approach — to deal with the source of the problems — in the late 1980s, which it announced in a memorandum it issued along with its budget. The water boards' attempts to clean up polluted water courses by burning sludge unintentionally led to further pollution (of both air and water) and thus was the main catalyst for this decision.

The shift from a sectoral to a theme-based approach was assisted by setting clear-cut targets and time frames for achievement, and the integrated source approach was formalized by the establishment of a 'target group management system'. The NEPP was characterized by a management approach to environmental problems involving:

- adoption of quantified (measurable) targets and time frames;
- the integration of environment into decision-making by all sectors of society;
- clear identification of responsibility for actions;
- creativity in the design and use of policy instruments;
- a commitment to long-term reshaping of social and economic structures; and
- recognition of the Netherlands' dependence on international cooperation and action.

A comprehensive report, *Concern for Tomorrow* (RIVM, 1989), described the state of the environment in the Netherlands and grouped the main environmental impacts (problems and pollutants) according to the economic activities responsible for them. It further categorized impacts at five different levels or scales of occurrence: local, regional, fluvial (river systems), continental (trans-boundary and marine) and global. 'The scale of impact had implications for the complexity of the source-to-impact chain, the time needed to require environmental quality and the level of government action required to deal with it' (VROM, 1993b). This analysis was incorporated into the NEPP's analysis of environmental problems into the eight environmental themes listed above. Each theme set quality objectives to meet an overall goal of 'sustainability by the year 2010 (or 2000)'.

The overall theme objectives of the NEPP were further broken down into numerous reduction targets for specified substances and waste streams. Responsibility for meeting these emission targets was with target groups — representing key groups of polluters in Dutch society, for example, industrial, agricultural and construction sectors.

The NEPP was prepared as a result of a process later laid down in the 1993 Environmental Management Act — integrated legislation on the environment which replaced previous sector-based laws. This required the NEPP to be revised every four years and extended it to the regional and municipal levels. NEPP was a very important tool linking sectoral policy developments. It was an 'umbrella plan' paralleled by associated four-yearly sector reports (on, for example, energy conservation, the state of water works, and the strategic plan for the development of infrastructure), each of which contained a 'heavy element' of environmental policies. The NEPP was formally agreed (signed) by five key ministries (Agriculture, Economic Affairs, Energy, Transportation, and Water Works). The NEPP's overall goal was sustainability by the year 2010. Various instruments were subsequently adopted to implement the NEPP (Box 14.1).

In 1989, the government fell, and the incoming minister wanted to make additions to the NEPP on CO_2 and NO_x. These were incorporated in a revised version in 1990 — NEPP+. The process took about five months.

Work started on preparing the second *National Environmental Policy Plan* (NEPP2) in mid-1992. It was published in September 1993 and approved in 1994. As early as 1992, two extensive reviews had been undertaken (VROM, 1994a):

- the National Institute of Public Health and Environmental Protection (RIVM) prepared Environment Outlook 3 (RIVM, 1991) — a study that reported on the current state of the environment and, given full implementation of adopted environmental policies, made forecasts of environmental quality in the years 2000 and 2010. The report therefore provided a measure of progress against environmental targets established by NEPP.
- The Ministry for Environment carried out a major evaluation of the policies developed by all the departments within it. Other sectoral ministries carried out reviews of their own policies where appropriate. The objective was to identify strengths and weaknesses at every stage of the policy process, from its first concepts through to its instruments, actors and enforcement. The results of the evaluation were then analysed in the light of Environment Outlook 3. This lengthy process pointed to the elements of NEPP that should be retained or strengthened and the new measures that should be adopted in NEPP2.

The *OECD Environmental Performance Review of the Netherlands* (OECD, 1995b) succinctly describes the aims of NEPP2 (see Box 14.2). The NEPP2 covers the planning period 1995–8 with a continuing overall goal of sustainable development within one generation.

While the first NEPP 'set out the government's environmental agenda and created the momentum for many other groups in society to develop their own plans and programmes', the function of NEPP2 was to 'follow through these many initiatives and to ensure that their objectives are realized and NEPP targets are met' (VROM, 1994a). Three core elements underpinned the strategy of NEPP2:

Box 14.1 KEY INSTRUMENTS IN THE IMPLEMENTATION OF THE NEPP, 1989

Direct regulation: *general legislative framework; permitting system for industry; environmental impact assessment; environmental quality standards.*

Voluntary agreements: *target group approach — covenants and action plans; codes of conduct.*

Environmental reporting: *public right to know; reporting requirements under industry self-regulation (covenants); corporate environmental management systems; enforcement.*

Environmental technology: *knowledge — technology assessment and communication; development — subsidies for research and technology sharing; application — demonstration projects and accelerated depreciation.*

Financial instruments: *financing charges of local government (waste, sewerage, water pollution charges); environmental taxes (fossil fuels, groundwater and waste); price signals to consumers (deposit refunds, tax incentives on clean products, high duty on petrol).*

Social instruments: *subsidies for environmental and other organizations and projects; environmental education in schools; intensive public information campaigns; provision of facilities to encourage behavioural change.*

- *Strengthening the implementation framework:* broader range of instruments; quality and effectiveness of instrument mix; and greater cooperation and integration between all actors.
- *Introducing additional measures:* new measures where targets cannot be met with existing policy alone (predominantly concerning CO_2 and NO_x).
- *Working towards sustainable patterns of production and consumption:* integrated life cycle management in industry; environmentally sound consumer behaviour; purchasing decisions and lifestyle choices.

Many activities are now focusing on the next steps that have to be taken to develop longer-term environmental policies in the Netherlands, for example, research on sustainable economic structures, lifestyles and sustainable consumption. In November 1995, the Minister of Environment announced in Parliament that the third *National Environmental Policy Plan* would be published in 1997. In contrast to NEPP2, NEPP3 will be focused on the relation between physical planning and economy. The contents of NEPP3 will also differ from the NEPP and the NEPP2. Furthermore, two policy documents will be published: one on environment and space (at the end of 1996) and one on the economy and environment (in early 1997). These two documents will provide the strategy for the integration of environmental policy with other policies on physical planning and socioeconomics.

Box 14.2 AIMS OF NEPP2, 1993

The NEPP2 aims to promote an environmental protection policy oriented towards sustainable development, taking into account the assimilative capacities of the media. This planning effort responds to several overall objectives:

- *to give perspective and structure to environmental policy;*
- *to integrate the activities of all ministries concerned with environmental management;*
- *to involve the public and target groups, and specify their undertakings; and*
- *to promote changes in production and consumption.*

In the effort to reach the goal of sustainable development by 2010, i e within one generation, each objective has led to the setting of more specific goals for environmental quality and pollution reduction, which are further broken down according to flows of substances. NEPP2 identifies the territorial contexts within which the various problems exist, and transcends the traditional problem categories drawn up according to media. It also identifies eight main themes and nine target groups:

Themes	*Target Groups*
Climate change	*Agriculture*
Acidification	*Traffic and transport*
Eutrophication	*Industry and refineries*
Dispersion of toxic substances	*Gas and electricity supply*
Disposal of waste	*Building trade*
Disturbance	*Consumers and retail trade*
Squandering of resources	*Research and education*
	Societal organizations

The plan sets out the conditions for integration of sectoral policies with environmental objectives. It recommends participation by communities and agencies, specifies action to be taken, identifies the necessary human and financial resources, and sets goals for businesses and municipalities.

Environmental planning is presented as a continuing process, including:

- *a report on the state of the environment (the 3rd edition of* Concerns for Tomorrow, *published in 1994, covers 1993–2015);*
- *a four-year plan; and*
- *an annual rolling three-yearly environmental programme, which accompanies the budget. It is based on the status of execution of the plan and specifies future action.*

In the light of the amount of progress of NEPP, NEPP2 recognizes that the measures contained in its predecessor will not enable some objectives to be attained. It draws up a balance sheet showing results so far, and demonstrates that diffuse target groups (consumers, transport users, businesses, small and medium-sized industries) are the most difficult to reach. It confirms the quantitative objectives of NEPP and proposes further action, while re-evaluating the instruments to be used and emphasizing social instruments. NEPP2 is oriented towards implementation and improved efficiency, including that of environmental monitoring and law enforcement. Never-

theless, evaluation of obstacles and delays does not appear to have been sufficient. The plans are instruments for integration of environmental concerns into sectoral policies: they take into account economic factors and request ministries to integrate environmental matters into their operations.

Source: OECD, 1995b.

Prime Motivation and Getting Going

The main motivation behind the NEPP and subsequent plans was to bring about a shift from a sectoral to a theme-based approach to environmental planning and management. In this respect, the Brundtland report (WCED, 1987) was very influential and 'provided a strong tail wind'. Also, public opinion was strongly in favour of the government playing a more active role on the environment. This had come about for various reasons, including the Chernobyl accident in 1986, the coming to light in the 1980s of domestic scandals concerning soil pollution in the 1970s, and the high public expenditure needed to deal with the pollution created by attempts to clean up the water (see previous section). In the 1970s and 1980s it was possible to smell polluted air and to see waste heaps grow.

The Cabinet, which very much wanted to 'take a lead', established the NEPP process, but there was also strong leadership from certain key individuals in VROM, who saw the need for NEPP, were not 'afraid of other influential ministries' and who 'made possible the enormous jump forward' represented by NEPP.

Focus

According to VROM staff interviewed, 'the NEPP, NEPP+ and NEPP2 were part of a periodically revised environmental strategy in a sustainable development context'. The main aim of NEPP was to 'get the sanitation process underway as quickly as possible — to deal with the pollution problem'. There was a very strong will to start the process of sustainable development planning, linking economic and environmental planning for a sustainable society. NEPP2 endeavoured to integrate environmental, economic and social perspectives. A research programme on economically sustainable structures and greening the taxation system was initiated. The environmental measures and actions laid out in NEPP2 are linked directly to the budget process in Parliament.

Organization and Management

The NEPP process was jointly coordinated by the Minister of Environment and the Director-General of VROM, and other key ministries were also deeply involved. The process was guided by a strong steering committee comprising the directors-general of all the involved government ministries and, in addition, a representative from the National Institute of Public Health and Environmental Protection (RIVM) — an independent institute heavily funded by government. The committee met every six weeks to two months and whenever otherwise needed. At first, the process was top-down with an *ad hoc* VROM project team acting as the secretariat and preparing NEPP,

NEPP+ and the initial phases of NEPP2 through an internal process. But the team became more of a coordinating body later in NEPP2, seeking to involve all interested parties. In NEPP3, the aim will be to get the staff of other ministries to play a direct role in the project team.

During NEPP2, the project team regularly consulted NGOs and target groups and tried to pick up their ideas through workshops. But their influence was limited since, in the view of VROM staff interviewed, 'they [NGOs] did not pick up the challenge to play a positive role — many saw their role as policing or checking rather than being creative and making a constructive input'. VROM expects this situation to change dramatically in NEPP3.

The NEPP reports were prepared by various sections of VROM, but there was very close consultation with other ministries, which commented on drafts and agreed on chapter contents. The project team undertook all editing, developed inputs, added new policies, and framed the overall document. Some 17 evaluation reports were prepared, each reviewing progress in individual sectors/areas. These were discussed with target groups in one-off working conferences when the overall NEPP outline was also explained. Five round-table meetings were organized separately with 'top' representatives of industry, consumer organizations, NGOs, employers' organizations and trade unions. The NEPP and NEPP2 processes took about two years and one year respectively, including six months for writing the plan documents.

Terms of Reference

The initiation of NEPP was heavily guided by the then Prime Minister because it was a top political priority. Eventually, in 1989, the government fell because of environmental issues (related to plans to eliminate tax rebates for using private transport).

The terms of reference for the NEPP2 were set by the ministers of the five key ministries involved. At first it was difficult to get agreement on them, but after 'detailed discussions' and when an outline had been agreed for NEPP2, the matter was eventually resolved by Cabinet.

Participation

The NEPP was developed as an internal government exercise, through extensive consultation and negotiation with all affected government departments. There was no real involvement of the public, industry or NGOs. A number of implementation strategies were developed through negotiations with target groups and industry, and 18 voluntary agreements or covenants were concluded with industry. These mechanisms and other structures were used subsequently to enable industry and NGOs to reflect on issues during the development of NEPP2.

During NEPP2, a network of some 600 individuals was involved (writing and/or participating in meetings, conferences and round tables) — from across government departments, research institutions, target groups, NGOs and local government. NEPP2 was still very much a 'consultative' process, but the aim was to focus on persuading the target groups to participate in discussing what changes they should make and to participate in monitoring implementation. The trend is to move away from direct (top-down, command and control) economic and environmental management instruments, to more socially negotiated and participatory instruments such as voluntary covenants with target groups (which set out monitoring and evaluation systems). Covenants, agreed under NEPP2, were always the result of a participatory process with the target groups. Such negotiation and public discussion is a part of Dutch culture. There is a long tradition of cooperation between government and industrial associations or local authorities. Though covenants were essentially started as 'gentlemen's agreements', with a highly uncertain status and degree of enforceability, they are now generally standardized and formalized with regard to procedure and content. The government sees them as a way of expressing joint responsibility (see Box 14.3). About 100 covenants have been concluded between the government and the private sector in recent years. It is still too early to judge the real success of the voluntary agreements struck under NEPP2. The *OECD Environmental Performance Review of the Netherlands* (OECD, 1995b) concludes that:

> *Progress has been made in the implementation of such agreements: they are stimulating cost-effective actions, and serve as more or less binding substitutes for regulation in a number of areas. However, should the environment cease to be a major public concern, it is not certain that these instruments would lead to substantial results. They must be used in association with other instruments and with mechanisms of accountability.*

NGOs had considerable indirect influence on the NEPP and NEPP2 via their activities in promoting public campaigns. They also had direct influence through their participation in working groups and round-table meetings. VROM would like to see them abandoning their traditional confrontational attitudes and becoming partners and participants in the NEPP process. It is aimed to make NEPP3 'as reasonably participative as possible'.

NGO, Public and Political Reactions

The final drafts of NEPP2 were discussed with the target groups to determine the acceptability of proposed measures. In the end, there was broad consensus on the package, but strong opinions against some of the policies were held by industry and NGOs.

A wide variety of official government advisory boards, NGOs, trade unions and employers' organizations formally responded to NEPP2. Their reactions have been brought together in a booklet, which has had an influence on the implementation of NEPP2 and will influence the direction of NEPP3. The NGO response was variable — some applauded NEPP2; others called its publication a 'black day'.

Milieudefensie believes that while the government's aim in NEPP3 appears to be to distance itself from environmental management through direct regulation, from its own discussions with target groups (see section below headed Action Plan: Sustainable Netherlands) it is evident that the latter do, in fact, feel that the government still has a role to play in regulation.

Box 14.3 COVENANTS IN THE NETHERLANDS

Covenants are generally seen as complements to existing legislation rather than as alternatives; they have a special role in meeting NEPP targets. While authorities still prefer to use laws and regulations to exercise control, covenants are used to speed up environmental improvements pending legislation, if there are too many uncertainties regarding the content of legislation to be drafted, if government intervention is needed only temporarily, or if covenants are likely to be less costly in terms of implementation or enforcement. Some 26 environmental covenants have been signed between the government and industry, dealing, inter alia, *with products, packaging, waste, and emissions in general. Since the advent of the NEPP, the focus has changed from products and packaging to production of waste and emissions.*

The main requirements for covenants between industry and the central government were recently laid down in a provisional code *covering procedural arrangements (especially information to politicians) and the content of covenants (objectives, requirements, period of validity, consultation, monitoring of compliance, evaluation, and settlement of disputes). The legal status of the covenants is generally that of an agreement under private law. If need be, the authorities can turn to the civil courts for enforcement. Consultation started in 1990 on environmental policy guidelines for the construction industry. After three years of discussion, the government and the industry adopted key objectives and signed an Environmental Policy Plan:*

Covenant with the construction industry (policy lines and selected targets)

A Reduction in use of non-renewable raw materials:
- *2.5 per cent reduction by 2000 and 5 per cent by 2005 with respect to 1990.*

B Stimulation of reuse of raw materials:
- *Reuse of construction and demolition waste to rise from 60 per cent in 1990 to 90 per cent in 2000.*

C Reduction in volume and separate collection of construction and demolition waste:
- *5 per cent quantitative prevention of demolition waste in 2000.*
- *80 per cent of demolition operations to use selective demolition techniques and separate collection by 1996.*

D Stimulation of use of renewable resources:
- *Tropical hardwood to be used only from sustainable-managed forests from 1995.*
- *Use of non-tropical wood to increase by 20 per cent between 1990 and 1995.*

E Reduction of use of harmful materials and substances:
- *At least 50 per cent of paint used by construction industry to be low solvent paint by 1995.*
- *Emissions of polycyclic aromatic hydrocarbons (PAH) to be reduced by 50 per cent in 1995 with respect to 1990 levels.*

F Promotion of energy saving heating systems and water efficient installations in new and renovated buildings:
- *Energy consumption of buildings to be decreased by 8 per cent by 1995 with respect to 1989/90.*

- *Water efficient installations to be fitted in defined per cent of new and renovated buildings by 1995 (for example, 50 per cent with water saving shower heads).*

Considerable progress has already been achieved concerning some of the targets. For instance, recycling of construction waste has reached about 60 per cent..

(OECD, 1995b)

Key 'Assistance' Factors and Problems

The initiation and development of the NEPP was greatly helped by public pressure and support for government action on the environment. There was also considerable government commitment at the highest level. Senior VROM staff played a pivotal role in pursuing NEPP without fear of opposition from other ministries. There was also strong support in the Cabinet for focusing the implementation on reaching 'difficult' target groups.

NEPP targets were defined in technical terms, for example, emission reductions. The evaluation process following NEPP concluded that a greater effort was needed to reach and focus implementation on the 'more difficult' target groups. As a result, NEPP2 did focus on these groups and policies were defined particularly for consumers and retailers. VROM believes that NEPP3 needs to do more to effect change in consumption and production patterns.

A key difficulty has been to quantify the specific contributions of the many target groups and subgroups towards reaching particular targets, so as to determine which groups need to 'do more' and to translate this into policy.

Implementation presented NEPP2 with a major management problem, particularly the question of 'how to make others the co-owners of the plan'. A 'management plan and control system' were established, which included 'tools' to ensure implementation of all the action points, of which there were 200 in NEPP and 160 in NEPP2. Strong persuasion was required in order to organize the necessary 'commitment to implement'.

Conflict Resolution and Consensus-Building

Many of the difficulties encountered during the development of the NEPP were caused by an initial lack of information. Also, the various ministries and different levels within government tended to hold contradictory opinions about issues such as energy taxes. These problems were dealt with through a predesigned process of 'open planning', described fondly by VROM as 'endless talk'. At the first level, the NEPP project team would address the problems in discussions with civil servants. Any unresolved issues were then handled by an intra-governmental task force at directorial-level bilateral negotiations with other ministries. If this still failed to resolve an issue, it would be raised by the steering committee of the ministries' Directors-General. Failing this, the Cabinet would act as the final arbiter of any outstanding conflicts. Target groups were also consulted.

The open-planning process in NEPP2 was facilitated by a training course on negotiation (entitled 'implementation challenge'). This course, coordinated by VROM, aimed to give all government participants a feeling that 'win-win' solutions could be reached. Professional trainers assisted VROM in designing the two or three-day course and in its delivery. The course is being continued to assist all levels of interdepartmental negotiations.

Conflict and opinion resolution was confined to government, but NGOs played an important role by influencing government through public campaigns and changing public opinion. 'The final goal of NEPP2 was not full agreement. It was accepted that there would be differences of opinion. It was the job of the Cabinet to share the pain.' According to VROM, discussions were 'tense' to the last day when the Cabinet decided to approve the NEPP2 document. This brought a 'sense of relief' and a document 'everyone could live with', for it had emerged from the open-planning process.

In large measure, the NEPP was a government response to a feeling in society that something needed to be done 'to clean up the environment'. There was less public interest in the environment during the preparation of NEPP2, when there was more concern about issues such as the European Union, immigration and crime. But there was still a great deal of media attention and debate.

Regional, Provincial, Local, Convention-Related and Dependent Territory Strategies

The NEPP2 addresses international issues such as OECD activities, CSD agreements, EC and EU regulations. The actions in NEPP2 are linked to Agenda 21 commitments, and these are highlighted on each page of the document.

In November 1994, a conference was organized between representatives of VROM and more local levels of government (provinces, municipalities, water boards) to discuss how the latter could implement NEPP2. Each province has its own environmental policy plan as a means of responding to its duties and obligations under law (for example, checking company permits). Municipalities have collectively developed an action plan for their responsibilities to implement NEPP2. Water boards also have particular responsibilities to implement elements of it.

After UNCED, several of the larger towns and cities decided to set up their own Local Agenda 21s. While these are wholly independent of NEPP2, they do, however, seem to be highly participative. NEPP3 would be well advised to draw on and learn from the experiences of places like Leiden, and share the process with them.

A biodiversity action plan for the Netherlands is being written as a cooperative effort between VROM and the Ministry of Nature Conservation, Agriculture and Fisheries. The need for such a plan is identified in NEPP2. It will obviously influence NEPP3.

VROM has prepared a strategy for the Dutch Antilles, but as a completely separate exercise from the NEPP process.

Ecological Footprints and Transboundary Issues

NEPP2 addressed transboundary issues throughout. Following its publication, VROM began to consider ecological footprint issues in its work to develop a vision of natural resource management in the Netherlands but in a worldwide sense.

Parliamentary Process

The NEPP was endorsed by Cabinet and presented to Parliament. The Cabinet approved the NEPP2 unanimously and, subsequently, Parliament approved it with only minor changes after a one-day debate in March 1994.

Press Coverage

The publication of both the NEPP and NEPP2 attracted a lot of press coverage and presentations were made by the Minister of Environment. The NEPP was taken very seriously as it arose at a time when the environment was at the top of the political agenda. While the press and public remained enthusiastic and positive about NEPP2, it was seen as less of a new thing because it was a logical follow-up.

Action Plan: Sustainable Netherlands

Initiation and Time Perspective

Two years before UNCED, Milieudefensie began to prepare for UNCED by considering what sustainable development might mean for the Netherlands in a concrete way. Work on developing the *Action Plan: Sustainable Netherlands* (APSN) was initiated by Milieudefensie in early 1991. That year was also the twentieth anniversary of Milieudefensie and the APSN provided a major initiative to celebrate the occasion. Using the concept of 'environmental space' (see Box 14.4), the aim of the strategy was to examine what a sustainable Netherlands would look like in 2010 and what steps would be needed to achieve it (see Box 14.5). It is not really an action plan, but sets goals for action, though it does not define precisely (or in detail) how to implement them.

Other national Friends of the Earth (FoE) organizations in Europe indicated a desire to follow-up the APSN. This has led to the three-year 'Sustainable Europe' project (see Table 21.2).

Focus

The main focus was sustainable development, based on the central theme of 'environmental space' (see Box 14.4). There is complementarity between the studies of Milieudefensie and RIVM (1989). The new element was that APSN was based on inputs, particularly in terms of land use, energy, non-renewable materials, wood and water. There was no direct link with the NEPP process, but the APSN has the same time horizon — to solve large problems within one generation.

Organization and Management

The APSN was written by a team from Milieudefensie. For a few subjects, research was contracted out to consultants. Many individuals collaborated in the work. The document was presented to a one-day conference attended by 400 people from across Dutch society. Keynote speeches were made by senior representatives from the Federation of Trade Unions, the main employers' organization and by the Minister of Environment. There was a series of workshops at the conference on actions/issues.

Box 14.4 ENVIRONMENTAL SPACE

'Environmental space' is the main theme of the Action Plan: Sustainable Netherlands. *It is concerned with*

> *the space that the Earth (nature) provides for humans (and other species) to exploit. At the moment, the environmental space is diminishing because of encroachment on nature and the environment. Because of population growth, the space available per person is smaller. Sustainable development means living within the limited environmental space. The wealthier countries are therefore asked to make extra sacrifices because their intensive use of the environment does not allow for proper economic development of other countries. The term sustainable implies far more than cleaner production and the prohibition of dangerous substances. Not only will drastic technical and structural changes have to take place, but human culture will also have to undergo changes until we achieve a mentality that prefers quality to quantity and doesn't take continuing increases in consumption for granted.*

Source: Milieudefensie, 1992.

APSN was subsequently presented at UNCED to the NGO forum, with some distribution of summary versions to the official conference. Milieudefensie received many complimentary letters and many articles on APSN have been written.

Milieudefensie admits that APSN was somewhat 'shortsighted' in that it was aimed both to coincide with its twentieth anniversary and to be presented at UNCED, and then to be discussed to see what effect it had had. There was no clear strategy for how to deal with the plan afterwards, apart from using it as a background document and the basis for future work on the issues described in the report.

Participation

The aim of preparing APSN was to 'set out the concept of a sustainable Netherlands for Milieudefensie in the first instance, and then to involve others to see what they thought and how to implement it'. There was, therefore, no process of participation as such in the preparation of the document. It was written by a team. However, after publication, it was subject to discussion and debate at a one-day conference and at many other meetings.

NGO, Public and Political Reactions

In the Netherlands, there is already a very high level of general environmental awareness — the public has a positive attitude towards environmental protection. But most members of the public, as in all countries, have limited technical or scientific knowledge and or a limited overview of the problems. The message of APSN was 'complicated' for 'the man in the street' and for non-specialist journalists. Nevertheless,

> **Box 14.5 ACTION PLAN: SUSTAINABLE NETHERLANDS**
>
> *In Part I, an attempt is made to translate the broad concept of sustainabiliy into concrete terms applying to the situation in the Netherlands, but starting from a global perspective for all key resources except water which is considered a regional resource.*
>
> *Based on the concept of 'environmental space' (or 'ecoscope': resources as globally available, used in a sustainable fashion, i e in a way that ensures that they will continue to be available in the future), a rough calculation of the global environmental space was made for the key resources of energy, wood, water, raw materials and arable land. Dividing the world environmental space by the number of world citizens produces the environmental space available per person, and consequently per Dutch citizen. The situation in 2010 (when the world population is approximately seven billion) was chosen as a target.*
>
> *Part II looks at the effects the calculations will have on consumption in the Netherlands. Sustainable consumption is defined by the main elements of daily life (for example, households and the home — including water use, household appliances and textiles; agriculture and food; recreation and leisure).*
>
> *Part III considers how to achieve a sustainable Netherlands and the contributions necessary from the main actors. International organizations, government, business, industry and the consumer are treated separately. Important issues requiring further research and debate are outlined.*
>
> *The main conclusion is that seven billion people could have a lifestyle similar to the present Dutch lifestyle; but with a lower consumption of meat and car/air transport.*
>
> Source: Milieudefensie, 1992

there was good coverage and discussion in the press and APSN was received well, especially by the more interested audience. There was much interest, both domestically and internationally. However, as expected, there were some critical comments. The Minister of the Environment was publicly enthusiastic about the report and praised it as a positive approach.

The APSN has not been subject to any official process or parliamentary consideration. The official government advisory committee on the environment has recently presented advice to the Minister of Environment on the concept of environmental space, drawing ideas from APSN. Although the APSN has had considerable influence in the Netherlands, it is difficult to measure.

Key 'Assistance' Factors, Problems and Issues

As with the NEPP, the high degree of public support for dealing with environmental issues in the Netherlands was helpful. The APSN built on the concept of the utilization of 'environmental space' introduced by Professor J B Opschoor and based on the earlier work of Paul Ehrlich and others. The most debated issue and the key element in the APSN was the global redistribution of resources. There is no consensus yet on the concept of environmental space and a sustainable Netherlands. But there is considerable acceptance of the need to set final limits to resource use. 'But people have not yet realized that, as a consequence, this poses a distributional problem that needs to be tackled. Some people deny there is a need to limit resource use — maintaining that technical solutions can be found.'

Chapter 15 NEW ZEALAND

Barry Dalal-Clayton and Barry Sadler

The key national strategy or green planning processes in New Zealand are the *Resource Management Act* (1991) and the *Environment 2010 Strategy* (1995). A state of the environment report is being prepared (due February 1996). Furthermore, a Sustainable Land Management Strategy and a Sustainable Water Management Strategy are being developed to give focus and further direction to the sustainable environmental management agenda (see section below on focus). A National Conservation Strategy was developed, but it was not implemented and 'sat on the shelf'. However, this strategy, the Brundtland report (WCED, 1987), and internal work in the Ministry for the Environment (MfE), have been ingredients in elaborating the concept of sustainable development within New Zealand.

Other supporting initiatives under development include:

- a framework for national environmental quality standards and guidelines for standards;
- a framework for monitoring the *Resource Management Act* and also for monitoring environmental quality; and
- a framework and process for comparative risk assessment and priority setting.

The Resource Management Act and the Environment 2010 Strategy

Initiation and Time Perspectives

The MfE is responsible for both the *Resource Management Act* (RMA) and the Environment 2010 Strategy. Each was started following recommendations made by the MfE to, and approved (after modification) by, the Cabinet.

Preparation of the RMA commenced in 1988 and was driven by the Cabinet Ad Hoc Committee on Reform of Local Government and Resource Management Statutes. Development of the Environment 2010 Strategy was started in 1994. The aim is that it should be subjected to a rolling review and revision process every four (or possibly five) years. It may be linked with the four-yearly state of the environment reports launched in 1995.

Prime Motivation and Getting Going

(a) Resource Management Act The RMA had its origins in the mid-1980s when the government was engaged in dramatic economic reform (both macro and micro) (see Box 15.1) driven by the Treasury which had a focused ideology that the 'way forward' was through more emphasis on markets and less on government controls (for example, institutional reform and deregulation). There were also problems concerning the functioning of government institutions responsible for environmental management. There was considerable concern about the inequities in the way that environmental management operated across different sectors, particularly from environmental and industry groups. In effect, the situation was a 'mess', with overlapping and conflicting responsibilities:

> *The reform process [repeal of the Town and Country Planning Act 1977 and enactment of the Resource Management Act 1991] was not only a rationalization of existing, admittedly often overlapping and contradictory, resource legislation, but also a deliberate move to limit the role of statutory planning in resource allocation decision-making. The wider socioeconomic objectives of the former legislation were viewed as unnecessary and undesirable interventions in the functioning of the market allocation mechanism and were removed.*
>
> (Grundy, 1993)

In 1984, the Labour Party replaced the National Party in government. While its economic policies were unclear, it had made public commitments on the environment which subsequently led to the establishment of the Ministry for the Environment (MfE). The new ministry staff realized the need for new legislation to deal with the institutional problems and, in 1988, proposed the development of the RMA. The proposal found resonance with the then Minister of Environment (Geoffrey Palmer) — a lawyer with a keen interest in law reform — and was endorsed by Cabinet with some modifications.

(b) Environment 2010 Strategy Development of this strategy, which started in 1994, was catalysed by Roger Blakely, then Secretary for the Environment. The aim is to set out a broader vision following the 'micro' reforms introduced by the RMA (logically, this broader strategy should have preceded the RMA reforms). The strategy is seen as a 'capstone' document with other components reinforcing and drawing from its agenda. A package of other initiatives is being worked on to take forward current thinking and fill outstanding gaps. For example, at present, there is no coordinated system for monitoring and measuring environmental change and its implications, although one is being developed. While it is difficult to define what is sustainable, there is a strong hope that, in the short- and medium-term, work arising from the Environment 2010 Strategy will give people a much better idea on what is unsustainable in New Zealand.

The present Minister of the Environment is involved closely in promoting the strategy as a positive agenda for the current and future governments.

Box 15.1 *SOME ECONOMIC REFORMS AND RELATED ISSUES IN NEW ZEALAND*

The New Zealand government has undertaken a series of radical economic policy reforms, and continues to do so. Examples of economic rationalization policies with far-reaching implications (both positive and negative) for sustainable development include:

- Privatization *Crown research institutes now operate on a fee for service and full cost recovery basis. Their initial operating assets included an environmental database and knowledge, which is now effectively proprietorial rather than public information.*
- Removing subsidies *This has had dramatic environmental effects. In agriculture, for example, there is no longer any incentive to utilize marginal country or overuse fertilisers and other inputs.*
- Polluter pays *In principle, environmental and social damages are now expected to be internalized in user costs. However, in practice, this may not necessarily be the case. An interesting exercise currently underway is looking at transport pricing where, in effect, road users are presently subsidized — in the sense that they do not pay the full costs of congestion, air pollution and accidents. But there remains the political issue of whether and how these costs are allocated among and assumed by road users (i e commercial vehicles and individual motorists). However, it is unlikely — from both a consumer and electoral point of view — that people will be prepared to accept the costs of the 'whole package' (although it is not yet known what it would cost). But people may be willing to 'sign up' to parts of it. In fact, there is a growing awareness of the need for emission controls on vehicles and these are considered to be potentially achievable.*

Focus

(a) Resource Management Act Environmental outcomes were the key focus for the resource management law reforms (which led to the RMA), but economic rationalization provided the policy framework within which the reforms were developed. The MfE relies primarily on the RMA to initiate a transition to sustainable development. The RMA is concerned with dealing with environmental externalities and/or allocating access to commonly managed (public) resources such as water, coastal space (including coastal and estuarine water) and use of air in terms of discharges affecting air quality. It established clear environmental responsibilities and requirements for assessment and planning. Every local authority and the national government is required to account for the effects of development, and must monitor the state of the environment with a view to adjusting activities accordingly (for example, ensuring that future applications for resource consents do not exceed environmental standards and objectives). New Zealand has avoided a 'centralized' approach to planning for sectors (for example, energy, transport) (as in the Netherlands), and such functions are devolved to local authorities. The RMA represents a very big 'stick' — indeed, it is probably the most comprehensive 'big stick' in the world in that it sets integrated 'bottom lines' which operate across all environmental media. But MfE does not have the resources to offer the number of direct financial 'carrots' that some other countries provide. But it does work with mechanisms such as voluntary agreements and uses its limited resources to maximize the 'carrot' dimension.

The intent of the resource management law reform was to 'reject the proposition that such law should be concerned with the planning and control of economic and social activity for any other purpose than to ensure the adverse effects of this activity on the environment were avoided, remedied, or mitigated' (Grundy, 1993, citing Gow, 1991). The approach to the business sector is largely one of cooperative partnership, backed by information and quiet advocacy, for example, promoting clean production.

The RMA rationalized existing legislation with an accompanying intent to limit intervention in favour of market processes. It also set out to promote improved environmental outcomes as a result of resource use and to foster the sustainable use of natural and physical resources — a response to increasing awareness (both locally and internationally) that many existing resource uses were unsustainable and were damaging the environment. Thus,

> *sustainable management is embodied in statute in the* Resource Management Act, *as the overriding purpose for guiding the use, development and protection of natural and physical resources. It is to be, henceforth, the guiding principle to all regional, district and national planning.*
>
> (Grundy, 1993, p5)

Apart from some Department of Defence activities and conservation within the public estate, which is dealt with under the 1987 Conservation Act, the entire government and private sector are now bound by the RMA.

Many statutory planning decisions are not, and never have been, backed by systematic monitoring to determine whether and how these make a difference. In short, while there is a far-reaching, information-rich institutional system for environmental management, there is no framework for drawing together the information and evaluating trends and cumulative changes. Though no documentation is yet available, a number of activities are responding to this deficiency. These include:

(i) *A State of the Environment Report*, which is currently under preparation and will cover key dimensions (like air and water) and sector-by-sector environmental impacts. The aim is to repeat the exercise every four years. An important supplementary activity is the development of indicators (for example, for ecological health).

(ii) A *Monitoring Framework*, which is being designed to set out key requirements for monitoring under the RMA. It will identify key objectives, protocols and methodologies. The thrust will be towards gaining a composite national measure of (a) environmental outcomes, and therefore of progress towards (or away from) sustainable environmental management — and so to help identify if interventions under the act are effective and achieving their stated purpose, and (b) the cost-effectiveness of the RMA systems.

Under the RMA, the environment is defined not only in objective, biophysical terms, but also includes socio-cultural values. Much of what constitutes environmental quality or sustainability (at least within certain limits) is culturally determined (i e by what communities want of and from their environment). This is most evident when Maori and European views of land, water and other resources are contrasted (see Box 15.2), but it also reflects other value axes (for example, urban–rural).

Box 15.2 *MAORI CLAIMS*

New Zealand has set a target date of 2000 to settle Maori land and other claims within a fiscal envelope of NZ$ 1 billion — which the Maoris have rejected (although they have not rejected the government's laudable and genuine approach). It appears that this will be a long-term, continuing process of negotiation, rather than a one-time, once and for all agreement. The MfE itself maintains a small Maori Secretariat to ensure that views and values of indigenous people are integrated within environmental policy and management decisions.

(iii) *National Environmental Quality Standards.* New Zealand has moved from a best-available technology basis to an effects-quality perspective on environmental management. The development of clear, quantitative environmental standards (for example, the capacity of air and water to assimilate industrial emissions) is a key to the proposed approach. At present, the focus of the work in this area is on a framework for clarifying, defining and prioritizing standards that reflect New Zealand's distinct biogeography, ecological history and particular 'receiving environment' media (see Box 15.3).

(iv) *Comparative Risk Analysis* work is about to begin, with the intent to assist people to make informed choices about significant issues, based on a better appreciation of knowledge about the risks and uncertainties, and the likely consequences of not addressing particular issues.

(b) Environment 2010 Strategy This is the government's strategy for the environment. The government takes the view that 'it is not its job to tell the people what sustainable development is — that is for the people to decide, but within the RMA and other relevant frameworks such as biosecurity legislation and hazardous substances and new organisms legislation'. However, the strategy does provide guidance on what the government sees as the important environmental issues and how they can be addressed.

The overall aim of the strategy is to improve environmental outcomes within the framework of a market economy and a cohesive society. While the strategy does not offer a systematic assessment of risks and priorities for action, it does provide a foundation for priority setting and ongoing review. It integrates the government's approach to environmental issues with its social and economic values. In this sense, the document is political — the strategy clarifies the aims of the government in terms of environmental goals and sets out an agenda for action.

Box 15.3 ECOLOGICAL CHARACTERISTICS OF NEW ZEALAND

New Zealand is characterized by several unique features:

- *as an isolated island state, New Zealand has had a long period of separate ecosystem evolution, followed by a relatively short era of European settlement;*
- *the latter is characterized by a range of imported agricultural practices and exotic biota that, in many cases, have drastically altered processes of ecological succession;*
- *intensive pastoral agriculture (for example, dairying) imposes very high pollutant loading on natural systems (for example, waste, nitrogen and phosphate run-off); and*
- *where animal waste enters water or where it has intensive effects on soil and land systems, its ecological impact is significant and pervasive (equivalent to a population of 100 million people), but it remains hidden from public view by a 'clean, green' visual image of the rural landscape A lot of animal waste degrades on the pasture and this is immediately recycled with no adverse ecological impact.*

> *The Resource Management Act 1991 requires us to take account of the needs of future generations. It is the cumulative effects of the way that we live over time that will determine whether future generations inhabit a more or less sustainable world.*
>
> *Naturally, a different government would wish to state its own values. But pressing matters of environmental priority are likely to be recognized as such by successive governments. The* Environment 2010 Strategy *provides the framework for an agenda for action while accepting that, in a democracy, different values will from time to time be brought to bear on these priorities.*
>
> (Hon Simon Upton, Minister for the Environment, Foreword to Summary of *Environment 2010 Strategy*, NZMfE, 1995)

The vision for the New Zealand environment to 2010 'is a clean, healthy and unique environment, sustaining nature and people's needs and aspirations' (see Box 15.4). In working towards this vision, a number of principles are proposed in the strategy that 'will underpin the government's actions in seeking to minimize or resolve conflicts between environmental, economic and social objectives (NZMfE, 1995) (see Box 15.5).

The strategy outlines the responsibilities of various groups for a six-part environmental management agenda (Box 15.6). It also describes the current state of the environment in New Zealand in terms of its strengths, problems, opportunities and threats. It then sets out 11 priority issues for the New Zealand biophysical environment (Box 15.7).

The strategy may be regarded as the nearest equivalent to a sustainable development strategy. It establishes an integrated agenda, bringing together other policy proposals, initiatives and responses to Agenda 21.

Box 15.4 VISION OF NEW ZEALAND'S ENVIRONMENT 2010 STRATEGY

The vision is for an environment where:

- *the life-supporting capacity of air, water, soil and ecosystems is safeguarded;*
- *biological diversity and spectacular scenery is conserved;*
- *the basis is provided for sustainable development that meets the needs of present and future generations;*
- *people are able to meet their needs, especially for employment, food, clothing, shelter and education;*
- *it is safe and healthy;*
- *natural, renewable resources are not consumed faster than they can regenerate;*
- *the natural treasures or* taonga *of Maori are protected, and the cultural practices of Maori associated with the environment are provided for; and*
- *leisure and recreational opportunities are provided for those who enjoy the outdoors.*

Source: NZMfE, 1994b.

Box 15.5 PRINCIPLES FOR INTEGRATING ENVIRONMENT, SOCIETY AND ECONOMY

- *Sustainable management of natural and physical resources;*
- *Application of the precautionary principle where knowledge of potential adverse environmental effects is limited;*
- *Recognition of 'environmental bottom lines' (i e thresholds below which ecological systems suffer irreversible damage);*
- *Internalization of external environmental costs;*
- *Encouraging the use of property rights approaches that achieve sustainable environmental outcomes;*
- *Evaluating local and central government interventions to ensure that 'least-cost' policy tools are adopted and only when necessary;*
- *Ensuring that environmental and social goals are mutually supportive;*
- *Public utility infrastructure pricing should follow full cost pricing principles;*
- *Considering local, national and international dimensions of sustainable resource management when investing in research, science and technology;*
- *Defining the limits of resource use and substitution; and*
- *Protecting New Zealand's international competitiveness.*

Source: NZMfE, 1995.

(c) Response to Agenda 21 As it stands, Agenda 21 is not seen (by those interviewed) as a particularly helpful or clear statement on achieving sustainable development. Most government departments have analysed the significance of Agenda 21 for their policies

> ***Box 15.6 ENVIRONMENT 2010 STRATEGY — ENVIRONMENTAL MANAGEMENT AGENDA***
>
> *The* Environment 2010 Strategy *provides a six-part environmental management agenda designed to help achieve the vision of a 'clean, healthy and unique environment, sustaining nature and people's needs and aspirations'. It aims to:*
>
> - *integrate environmental, economic and social policy;*
> - *establish a coherent framework of law;*
> - *sharpen the policy tools;*
> - *build up the information base;*
> - *promote education for the environment; and*
> - *involve people in decision-making.*
>
> Source: NZMfE, 1995.

and activities but, in general, only broad implications have been drawn. Agenda 21 is seen as being most relevant for catalysing actions by the business sector and local authorities.

The role of business and industry in relation to Agenda 21 has been outlined in the report of a national conference of business leaders (NZMfE, 1993b). A particularly notable initiative is cleaner production demonstration projects established through the MfE under a concept promoted by UNEP (NZMfE, 1993a). These projects are 'win-win' situations in which waste minimization efforts both reduce impacts and result in economic savings, and thereby confer some competitive advantage. On a wider front, a potentially important innovation is the Brand NZ trade marketing effort by government and private sectors. It requires companies to meet designated 'clean and green' total quality management standards for agricultural production from start to finish (for example, for meat production and packaging), in an attempt to win a greater share of international niche markets for 'environmentally acceptable produce'.

(d) Initial Emphasis The view is taken that, now that the broader policy and institutional innovations have been put in place, it is important to concentrate on more immediate and pragmatic aspects of sustainable environmental management. The initial emphasis will be on 'unsustainable development', notably understanding and clarifying notions of irreversible ecological change, significant impacts, and managing risk and uncertainty. Specifically, this will involve getting a better picture of:

(i) biophysical limits and thresholds;
(ii) the processes and activities that push against these; and
(iii) the needs and expectations of local communities (for example, what pressures they are likely to exert, and when trade-offs may need to be made).

Box 15.7 PRIORITY ISSUES AND ENVIRONMENTAL GOALS IN NEW ZEALAND

Eleven environmental goals have been identified in the Environment 2010 Strategy *that the government believes New Zealand can and should aspire to. The strategy recognizes that setting environmental goals at the national level can be somewhat arbitrary and that these goals 'will need to be refined in their application to particular situations and localities over time'. The goals are also indicative as opposed to providing a standard that can be met in every case. They will be pursued 'only to the extent that the government, the private sector and the community are prepared to commit resources to them'. They are not ranked in priority order:*

1. **Managing land resources** *to maintain and enhance the quality, productivity and life-supporting capacity of New Zealand soils, so that they can support a variety of viable land use options.*
2. **Managing New Zealand's water resources** *to manage the quality and quantity of surface water, groundwater, coastal and geothermal water so that it can meet the current and future needs of ecological systems, communities (including Maori), primary production and industry by: maintaining sufficient water in water bodies to meet these current and future needs; ensuring New Zealand's surface freshwaters and coastal waters are of a quality suitable to meet community needs such as swimming, fishing and shellfish gathering, and that aquatic life is not significantly affected by discharges and abstractions; restoring, and preventing further degradation of, groundwater quality and quantity; and preventing degradation of quality and flow of water resources that are identified as having national significance to New Zealanders for recreational, scenic, scientific or cultural reasons.*
3. **Air quality**: *to maintain air quality in parts of New Zealand that enjoy clean air, and to improve it in places where it has deteriorated.*
4. **Habitats and biodiversity**: *to protect indigenous habitats and biological diversity by: maintaining and enhancing the net area of New Zealand's remaining indigenous forests and enhancing the ecological integrity of other remaining indigenous ecosystems; and promoting the conservation and sustainable management of biological diversity so that the quality of indigenous and exotic ecosystems is maintained or enhanced to guard against extinction and permit adaptation to changing environmental conditions.*
5. **To manage pests, weeds and diseases** *by reducing the risks they pose, to levels consistent with New Zealand's established objectives for biological diversity of ecosystems, people's health, and biosecurity of the economy.*
6. **Sustainable fisheries** *to conserve and manage New Zealand's fisheries for the benefit of all New Zealanders by providing for sustainable utilization of fisheries resources, including commercial, recreational and Maori customary take.*
7. **Energy services** *to manage sustainably the environmental effects of producing and using energy services.*
8. **Transport services** *to manage the provision of transport services in a manner that minimizes adverse effects on the natural and physical environment and human health.*

9. **To manage waste, contaminated sites and hazardous substances** *by: managing waste to reduce risks to environmental quality and public health to levels that are widely agreed as being socially acceptable; cleaning up contaminated sites to reduce risk to the environment, people and the economy; and managing or preventing the harmful effects of hazardous substances in order to protect the environment and well-being of people and communities so as to enable the maximum net national benefit to be achieved.*
10. **Responding to the risk of climate change**: *to take precautionary actions to help stabilize atmospheric concentrations of greenhouse gases in order to reduce risk from global climate change and to meet New Zealand's commitments under the UN Framework Convention on Climate Change, including: to return net emissions of carbon dioxide to no more than their 1990 levels by the year 2000 (but aim for a reduction in net carbon dioxide emissions to 20 per cent below their 1990 levels by the year 2000 if this is cost-effective and will not harm New Zealand's trade) and to maintain them at this level thereafter; and to reduce net emissions of other greenhouse gases — particularly methane — by the year 2000 where possible, and maintain them at those levels thereafter.*
11. **Restoring the ozone layer**: *to help achieve full recovery of the ozone layer and constrain peak levels of ozone destruction by phasing out imports of ozone-depleting substances as quickly as possible and at rates no less than those agreed internationally, and by limiting, where practical, emissions of those substances that are imported.*

Source: NZMfE, 1995.

It is accepted that these also need to be set in the broader context of environmental trends and issues. Major environmental problems in New Zealand include ground and surface water contamination, and wide-scale erosion of marginal hill country. In many cases, environmental management has to take account of complex cultural-historical realities, for example, the combination of rabbit infestation and overgrazing in South Island that leads to close-cropping and eventual stripping of tussock grasslands.

(e) **Sustainable Land and Water Management Strategies** New Zealand has seen 150 years of agricultural development, and in recent years considerable efforts have been made to deal with the physical effects of agricultural development. Since 1941, there has been legislation dealing with soil and water control. Nevertheless, there is growing evidence that adverse impacts are increasing, with potentially serious consequences for resource productivity. The 1991 RMA aims to deal with the ecological effects and, together with other policy and institutional developments, it represents a more systematic integrated approach to sustainable resource management. The Sustainable Land Management Strategy (Box 15.8) and Sustainable Water Management Strategy (Box 15.9) are intended to take this approach further.

Organization and Management

Core responsibility for both the RMA and the *Environment 2010 Strategy* is with the Ministry for the Environment (MfE).

Box 15.8 NEW ZEALAND'S SUSTAINABLE LAND MANAGEMENT STRATEGY

The Sustainable Land Management Strategy (SLMS) is intended to pull together several strands of activity, including responsibilities of private land managers, information, education, communication, and monitoring under the RMA.

A sustainable management fund of NZ$ 5 million has been established. However, this also includes a range of issues beyond land management, and the fund is small in relation to the scale of problems.

Major land management issues in New Zealand include:

- *severe erosion of east coast forest, where massive landslides have occurred as a result of a combination of relatively arid conditions, concentrated run-off, large-scale conversion of indigenous forest to pasture, and shallow root systems;*
- *urban expansion into higher quality soils;*
- *ecological deterioration and weed infestation of South Island high country, caused by a combination of tussock burning, domestic grazing, and (exotic) feral species — notably rabbits; and*
- *tourism and recreation impacts on national parks of South Island becoming visible in high use/visitation areas — notably around Queenstown. These are likely to increase with the trend toward increasing tourism and 'high technology' backcountry activities (for example, jetboats and scenic flights).*

Many of these problems are complex, encompassing a range of interconnected issues. For example, a Sustainable Land Management Strategy for the South Island high country must address the changing economic structure, which has encouraged the substitution of tourism for farming; security of tenure (much of the land is owned by the Crown and is leased to farmers); individual responsibility for maintaining grazing carrying capacities; and institutional arrangements to secure maintenance of and public access to heritage and landscape values.

At present, the SLMS is an evolving document. It is expected that primarily it will identify the main issues and options for resolving them, and indicate the roles and responsibilities of the key players. This approach will complement and extend the planning and regulatory apparatus for sustainable resource management under the RMA, which sets what can and what cannot be done.

Looking ahead, implementation of the SLMS is likely to depend on cooperation of the interested parties. The prospects appear better in some areas than others. Where the problems are already acute (for example, the South Island hill country), farmers are already looking at how to work with government on tackling specific issues.

(a) **Resource Management Law Reform** (leading to the RMA) A four-person core group of selected experts (not representative) was established, including a lawyer, an economist, a Maori representative, and an MfE convenor (also an economist). This core group initiated a three-phased approach, based on substantive public information and input. This group reported to the Cabinet Ad Hoc Committee on Reform of Local Government and Resource Statutes, and devised a special stream of Maori consultation.

Box 15.9 NEW ZEALAND'S SUSTAINABLE WATER MANAGEMENT STRATEGY

Early problems of water management in New Zealand were often a question of too much water, for example, flooding and sedimentation. With increasing agri-industrialism in the postwar period, water quality issues developed. The regulatory arrangements and management by water catchment boards that preceded the RMA were reasonably effective in controlling point source pollution, and in allocating water use among competing interests. More recently, however, with the intensification of agricultural land use and farming practices, non-point sources of water pollution have become a serious problem. To date, the planning machinery has had limited success in dealing with this problem.

The Sustainable Water Management Strategy (SWMS) recognizes that existing policy and regulatory instruments do not work effectively to control such non-point sources of water pollution. These discharges are also difficult to monitor (as now required by consents under the RMA). The SWMS is evolving through (i) preparation of an issues paper, and (ii) consultation with key players (to date, the environmental lobby in New Zealand has targeted mining and forestry rather than agricultural activities, but this is now beginning to change).

The current phase of SWMS development is being taken forward by a small working group of 12 stakeholders. Following a scoping process, the group is attempting to negotiate a consensus strategy. It is expected that this will recognize and reflect both economic and political realities, and sustainability 'rules' and principles. Key components of strategy development are as follows:

- **Institutional structure.** *In New Zealand, Regional Councils are responsible for water management, and District Councils for land management. Agricultural practice and problems lie at the intersect of both regimes. The need is for either a transfer of responsibility or a cooperative mechanism for joint planning to ensure that land and water management policies are complementary.*
- **Information requirements.** *Water quality impacts are pervasive, have wide-ranging downstream effects, and spill over to other interests. However, the problem is not necessarily apparent to the farmers responsible. The need is to provide information and feedback that link causes and effects, and to change policy to generate 'on the ground' results. This requires significant research and data management, involving monitoring and evaluation of effects on water quality, and packaging information to make it understandable to farmers and others.*
- **Incentives for change.** *In essence, the SWMS involves a shift from land use zoning to an effects monitoring and impact management approach. Policy signals and rules reflect the first step in this direction. However, reliance on regulation alone is insufficient to deal with the issues at stake, and runs counter to the economic rationalization and market reforms being pursued by the government. For the longer term, major emphasis is being placed on developing incentives that reinforce the responsibility of individual farmers to minimize the impacts of their activities on water quality.*

Phase 1 initiated a debate on issues from zero base. Issues papers and public discussion documents were developed posing questions about the purpose of New Zealand's planning and resource management laws. A public consultation programme was undertaken around these papers involving public meetings, seminars, free phone-in facilities and written submissions. All comments received were entered into a database to prepare a profile of issues. Key words were developed to retrieve information for analysis. Responses were analysed by the core group of experts (chaired by the MfE) and by other commissioned experts. All papers submitted to government on the RMA highlighted where stakeholder views accorded or differed from proposals being made. This phase resulted in a recommendation that there was a case for having a statutory framework to deal with environmental and resource management issues, and the development of some principles, both of which were adopted by the government.

Phase 2 went on to look at options for achieving the recommendations. The Cabinet Ad Hoc Committee on Reform of Local Government and Resource Management Statutes decided that the focus should be to manage environmental externalities and set the task for phase 2 to develop appropriate mechanisms and devices. A similar consultative approach to phase 1 was adopted but focused more on what ways could be adopted to achieve the recommendations. The process included a special 'stream' for Maori consultation which involved meetings (traditional Maori *hui*) with tribal organizations throughout the country to explain the RMA process and to secure views and opinions. Recommendations were submitted to the Cabinet Ad Hoc Committee detailing proposals (for example, the focus should be on sustainable management). A decisions was taken that there should be a single *Resource Management Act* to achieve the agreed objectives determined at the end of phase 1.

Phase 3 involved intensive debate on the policy details and these were put into a legislative framework by legal drafters. The core group was broadened to a wider group in the MfE to steer the draft RMA through for submission to Cabinet for approval. The draft RMA was then introduced into Parliament in 1989 (but not passed).

A new National Party (Conservative) government was elected in 1990 and decided to review the proposed act. MfE staff were able to persuade the new minister (Simon Upton) not to open up the whole process but to focus on examining a few key issues that would benefit from further scrutiny, for example, definition of sustainable management. An external review group was appointed, headed by an independent lawyer, and elements of the proposed RMA were reviewed publicly through meetings and seminars. The revised RMA was reintroduced into Parliament in 1991 and was examined by a select committee, which introduced some further changes before the Parliament passed the Act in July 1991. It came into force in October 1991.

(b) **Environment 2010 Strategy** The strategy has been developed through a more truncated and 'top-down' process than the RMA. The idea of the strategy emerged in 1993 and the process was initiated in 1994 with a 'wise heads' consultation exercise. A draft document (known as the consultation document) was launched in October 1994 and public submissions were invited (NZMfE, 1994b). Public meetings were held to explain the initiative and proposals. The submissions were subsequently analysed and considered and a revised draft was circulated to all government departments, a reference group of industry, non-governmental organizations and local authority people, and external peer reviewers. The revised *Environment 2010 Strategy* was adopted by the New Zealand government in July 1995 and published two months later (NZMfE, 1995).

Terms of Reference

The Cabinet approved a proposal by the MfE to develop the RMA and set the terms of reference for it. The Minister for the Environment presented a proposal to the government Standing Committee on Environment and Industry, which endorsed draft terms of reference for the *Environment 2010 Strategy*.

Participation

In general, the RMA was a participative process (see section above on organization and management). NGOs were directly involved and some were actually funded to undertake commissioned work. The *Environment 2010 Strategy* was essentially prepared on a 'top-down' basis with some consultation. NGOs, the public, business and others were able to attend meetings and make submissions (written and by telephone), but were not 'formally involved'. Paradoxically, one of the goals contained in the draft strategy is to 'ensure that people have the opportunity for effective participation in decision-making that affects the environment' (NZMfE, 1995, p58)

The strategy also points out that under the *Resource Management Act*, 'central government and regional and territorial authorities are obliged to consult widely with the public, and specifically with *iwi* [Maori] in making policy decisions affecting the environment. Public participation is a part of the framework for establishing environmental policies in regional policy statements and regional and district plans. Processes for public participation should be both "user friendly" and efficient' (NZMfE, 1995, p58).

NGO, Public and Political Reactions

There was a positive response to the RMA when it was enacted and published, and environmentalists were generally unified in their support because, in its purpose, the new act embodied some fundamental ecological and environmental principles. But there was scepticism about the draft *Environment 2010 Strategy*, which environmentalists viewed as a government 'whitewash' presenting a government view with no societal consensus. Industry was somewhat neutral in its response.

Key 'Assistance' Factors, Problems and Issues

The development of the *Resource Management Act* was greatly helped by the 'sea change' which occurred following the fall of the National Party government of Prime Minister Muldoon in 1984 when there was a swell of public pressure in favour of environmental reform. But there were also several problems. The minerals industry at first adopted an obdurate and negative stance, and then realized that it was 'running against the tide' and changed course to support the process constructively. NGOs were (and still are) sceptical about implementation of the RMA and particularly about their ability to participate fully in the RMA process. In this context, they asked for legal aid and other means by which to participate. Before development of the RMA began, the main opposition in government was within the then Ministry of Works (responsible for water, soil, and town and country planning), which viewed the RMA as counter to its vested interests. This ministry was subsequently abolished and many staff were absorbed into the MfE. The RMA also had to deal with a number of key issues: the confusing 'mess' of overlapping and contradictory resource legislation and responsibilities among various agencies; the philosophy and idea of sustainable resource management (see Upton, 1995); and major rights and interests with respect to natural resources.

The *Environment 2010 Strategy* was facilitated by the publication in 1992 of the government's *Path to 2010* (Government of New Zealand, undated), which set its perspective on integrating environmental, economic and social policies and strategies. It has also been greatly helped by the fact that the current Minister of Environment (Simon Upton) has been very supportive and has strongly promoted the strategy initiative. The final version of the strategy was published in 1995, and problems that might arise with its further development and implementation are not yet known, for the first MfE annual stocktaking of its progress has yet to be completed.

Conflict Resolution and Consensus

The RMA core group was responsible for balancing, integrating and deciding between the different positions of interest groups. Thus, conflicts and differences were resolved by the MfE rather than through consensus-building. MfE staff interviewed are of the view that, in general, there is consensus on the RMA across government and society. It is too early to determine what consensus there is on the *Environment 2010 Strategy*.

Links to National Planning and Decision-Making

The RMA is concerned with the management of natural and physical resources. As such, it is related integrally to functions concerning national planning and decision-making, although it is not itself a planning instrument. The *Environment 2010 Strategy* document points out that the

> *annual government budget cycle provides an opportunity to consider the environment strategy in the broader context of the government's overall strategy and priorities.*
>
> *The budget cycle now includes a phase for establishing the government's strategic priorities in the short, medium and long term. These strategic priorities are 'bedded in' through budget appropriations, purchase agreements and strategic and key result areas in chief executive's performance agreements.*
>
> *Chief executives of government departments that have responsibilities in relation to the environment have been asked to take into account in their annual planning the goals of the* Environment 2010 Strategy *that are relevant to their responsibilities.*
>
> (NZMfE, 1995, p60)

The linkage between the annual planning cycle and the planned four yearly review of the *Environment 2010 Strategy* is shown in Figure 15.1.

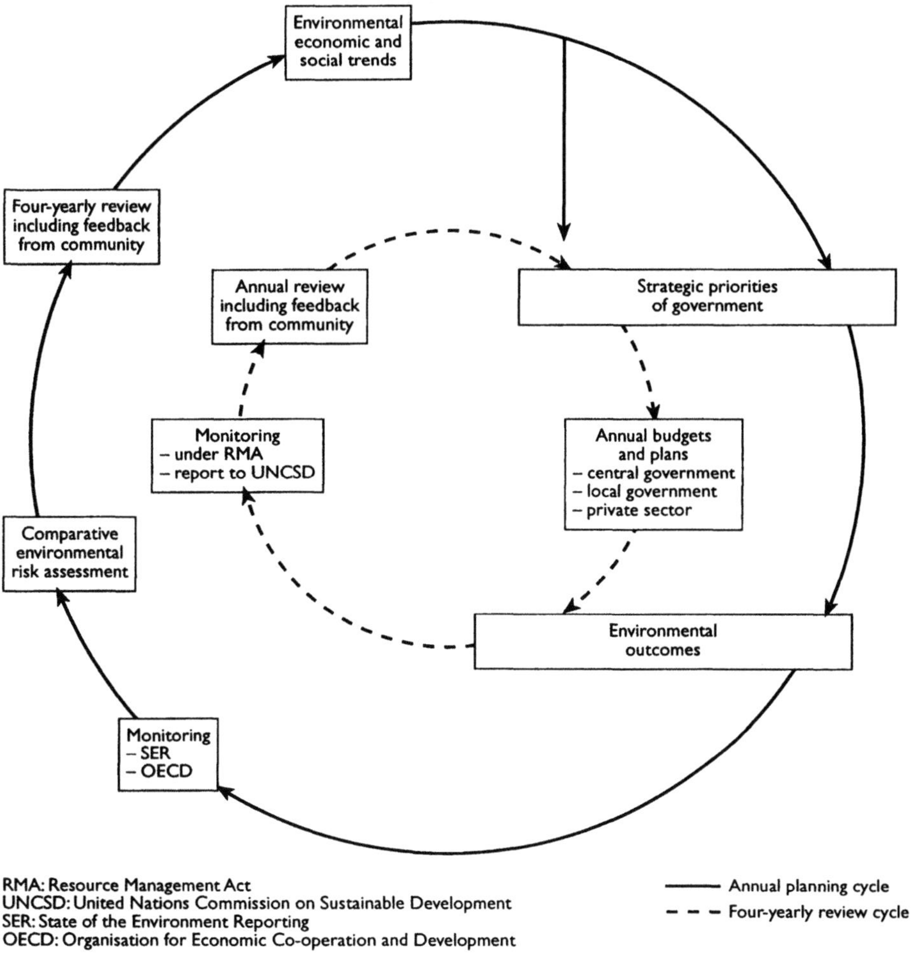

Figure 15.1 *Process of Setting Priorities and Monitoring Outcomes for New Zealand's* Environment 2010 Strategy

Source: NZMfE, 1995.

Local and Convention-Related Strategies

As mentioned above (p56), the MfE has set up a framework with which to help local authorities implement their own Local Agenda 21s; and has also given them a certain amount of technical assistance. A number of cities, such as Waitakere and Hamilton, have now developed their own local strategies.

New Zealand's Department of Conservation is developing a biodiversity strategy. This may be integrated into the RMA through the national policy statement (an administrative executive instrument under which the RMA can be revised). The issue of climate change is already subject to the RMA through a permitting system.

Ecological Footprints and Transboundary Issues

Neither the issue of ecological footprints nor trade are addressed explicitly in the RMA, but the theme of carrying capacity runs strongly through the act and could readily provide a vehicle to consider how concepts such as ecological footprints operate in the working of the RMA. The act is an enabling framework within a statutory system that sets some firm but descriptive environmental 'bottom lines'. 'The imperative of its purpose is very much concerned with ecology. It is, therefore, highly likely that ecological footprints will become the substance of quite a lot of its policy as developed, particularly by regional councils' (Gow, personal communication).

Greening the Mainstreams

The RMA has had a very broad influence. It set tough environmental 'bottom lines'. Industry behaviour has changed a great deal and it has been very keen to 'get things right' to deal with the 'threat' of a tough law and large fines. It has been 'greened' considerably and many industries are starting to develop their own codes of practice, while some sectors are developing their own strategies (for example, forestry, vegetable growers, pork, minerals and chemicals industries).

However, there remain problems with implementation. Some companies are trying to bully NGOs and objectors and thwart their using the appeals provisions of the Act (for example, threats to sue objectors).

Chapter 16 | NORWAY

A range of planning and policy instruments are considered together to be the equivalent of a green planning or sustainable development strategy process in Norway. They are listed in Box 16.1

Reports to Parliament on the Follow-Up to the Report of the Brundtland Commission (No 46: 1988–89), and on UNCED (No 13: 1992–93)

Initiation and Time Perspectives

(a) **Report No 46:1988–89** The preparation of the White Paper, *Report to the Storting No 46 (1988–9), Environment and Development: Programme for Norway's Follow-Up of the Report of the World Commission on Environment and Development* (NorMoE, 1989) took two years to complete. It was presented to Parliament on 28 April 1989. The process started in 1987, a few weeks before the Brundtland Commission's report (WCED, 1987) was published. It was initiated by a letter from the Prime Minister to all ministers requesting them and their colleagues to examine the WCED report and its recommendations, to identify what had already been done in Norway, and to say which of its elements were acceptable and could be implemented and which were not. Staff of the secretariat responsible for Report No 46 explained the WCED report in detail to senior civil servants in all the ministries. Following this, each ministry produced a list of actions and policies. These lists were reviewed by the secretariat, and questions as well as suggestions for action were presented to the ministries for further elaboration.

This report had no particular time perspective and did not set out to describe the shape of future Norwegian society and how to get there. Such 'visioning' is carried out by the Ministry of Finance's long-term plan produced every fourth year, in which all sectors of the society, the economy and employment are analysed. These plans look towards 2040 on some issues, and much further on demographic trends. Predictions of pollution levels, oil income and trade balances are made for well into the twenty-first century. The long-term plan contains a chapter on the environment by the MoE.

Report No 46 both brought together problems and policies concerning the environment in a single policy document, and set out the responsibilities of individual sectors for achieving environmental aims by appropriate means. The document places emphasis on cost-effective and economic measures, as appropriate, to reach the aims (see Box 16.2).

An operating mandate for the process was issued at the outset by the State Secretary

Box 16.1 *STRATEGY PROCESS IN NORWAY*

- Report to the Storting No 46 (1988–9), Environment and Development: Programme for Norway's Follow-Up of the Report of the World Commission on Environment and Development *(prepared by Ministry of Environment) (NorMoe, 1989) and published in two parts. Only Part 1 has been translated into English. Part 2 details the obligations of each ministry and contains a long chapter on economic and administrative instruments.*
- Report to the Storting No 13 (1992–3) on the UN Conference on Environment and Development, Rio de Janeiro *(NorMoE, 1992b).*
- Annual Environmental Policy Reports *to Parliament by the Minister of Environment.*
- Annual 'green budgets' *detailing the environmental measures taken by all ministries in their annual budget proposals (Report No 1 to the Storting each year).*
- *The national preparatory process for UNCED.*
- *The Department of Environment is now encouraging local communities to develop sustainable development plans rather than environmental ones. It is reviewing how best to provide financial support for this. The DoE is also trying to integrate its regulatory functions with incentives work.*

Other strategies being prepared include:

- *An integrated biodiversity strategy and action plan, coordinated by the Ministry of Environment. Nine ministries, local authorities and stakeholders are involved.*
- *A climate change plan for Norway—prepared by civil servants only within certain ministries.*
- *As announced in Report No 46:1988–9, the Ministry of Environment is financing the post of one executive officer in each local authority to take on the responsibility for environmental issues and to assist in the preparation of local environmental plans. A number of local communities are preparing independent Agenda 21s and eco-community plans through a broad consultation process.*

The general planning process in Norway, as in many other countries, includes public hearings and deals with local amenity and planning issues.

of the Ministry of Environment. This established a group, comprising the State Secretaries of all the ministries, which met regularly (approximately every three weeks) to review ideas emerging from individual ministries. A number of other specialist groups were set up to look at particular issues, for example economic instruments such as tax and how to use budget systems.

The WCED report was distributed to all 434 local authorities in Norway and 1000 copies were circulated to NGOs, with a request for comments on the report's implication for Norway. One-third of the authorities responded and their comments are

Box 16.2 ECONOMIC CONSIDERATIONS IN NORWAY'S REPORT NO 46:1988–9 TO THE STORTING

New results from economic models to forecast pollution levels, the costs and the effects of some pollution control policies were expected at about the same time as our green plan would be finished. This, together with a general uneasiness in relation to the costs of environmental measures to the economy, led to a tight coupling between the pollution control chapters of the plan and the budgetary process. As a consequence, only the less numerate chapters of the report could be negotiated in the familiar ways, balancing environmental interests against agriculture, discussing the need for constraints on fisheries and so on. The chapters touching on industry and trade, however, required a new approach.

In formulating aims in the past, the environment authorities have considered thresholds such as carrying capacity and health-related limits. In cases with severe health risks and a linear relation between emission and damage, minimum emissions were sought. This led to the use of concepts such as 'best available technology'.

Making plans for the entire environmental field, however, concerned the entire economy, and costs and benefits had to be balanced. This called for background material from other arms of government; difficult value judgements and questions of time perspective were added to the scientific uncertainty that we were used to handling.

Norway has a very open economy. If our main trading partners do not set stricter environmental rules at the same time as we do, our competitive position will be adversely affected. In addition, separate standards for tradable goods are difficult to implement in a situation with a free flow of trade.

Hence, except for CO_2, no new major aims were elaborated in Report No 46. The aims set out were the old aims on NO_x and SO_2 of the LRTAP *[Long Range Transboundary Air Pollution] convention, our aims based on the Montreal Protocol and the 'North Sea aims' on N, P and toxic metals, already agreed. Innovations here must be regional or global.*

In elaborating the implementation of these aims, two different approaches were discussed.

The 'economist' tradition avoided numerical aims in favour of general economic measures to be adjusted as the need arises. Specific commitments sector by sector are unnecessary since cost assessments made by each economic actor will automatically distribute action in the most cost efficient way.

The 'administrative' and 'engineering' tradition favoured technical standards and specific emission permits. These could act faster, more directly and be adjusted to local needs.

Both approaches could have been used to derive a 'Dutch model' of intermediate national aims for selected years, as well as calculations of economic effects. Intermediate aims were not set, however, and the economic calculations were not published together with the environmental plan.

The main new political commitments were to stabilize CO_2 emissions by the year 2000 at the 1989 level and to provide new and additional funding of up to 0.1 per cent

> *of the GNI [gross national income] for global climate measures abroad, provided that other countries followed suit. Although the economical effects had been analysed and found to be acceptable, the main decision to adopt the CO_2 aim, was a political choice.*
> (Hofseth, 1993)

summarized in Report No 46. Two innovations were introduced at the beginning of preparations for Report No 46: (i) an annual statement to Parliament on the state of the environment by the Minister of Environment, followed by broad debate and (ii) preparation of an annual 'green budget'.

(i) Under this new initiative, each Spring, the Minister of Environment now makes a speech to Parliament accompanied by a published statement presenting issues and supporting data. It contains no budgetary recommendations. This vehicle allows problems/issues to be highlighted within Parliament, and could effectively 'move the monopoly on initiating policy away from Cabinet into Parliament'. However, Parliament has not really grasped the opportunity to use this to debate the questions in their totality — since there is a tendency to use the occasion to focus attention on one or two 'burning issues' of the day, where more short-term political points can be scored.

(ii) Each year, the government's overall budget is presented to Parliament in early October. The budgets for all ministries are consolidated within one document. The idea of a separate 'green budget' — drawing together and making visible what budget flows in all ministries are directed to the environment — was first discussed within the Prime Minister's Office in August 1987. Cabinet approved the approach and there was detailed discussion with individual ministries. The Ministry of Finance wrote to all ministries providing guidance on preparing such a green budget within a framework for expenditure increases/decreases. The green budget is a published extract of the main budget but is amplified and illustrated as a separate document, with the aim of reinforcing the environmental responsibility of each ministry. The portions of ministry budgets that serve the environment are divided into three parts: money directed specifically towards environmental improvement, money spent with multiple aims, and funds for other purposes but which have incidentally positive environmental effects. However, there is no measure of the environmental efficiency of such measures The green budget was first introduced as part of the 1989 budget, presented to Parliament in October 1988. The aim was to make this green budget document a 'steering tool' for Parliament concerning the environment. Unfortunately, Parliament has not really used it.

Usually, Cabinet prepares policy documents on specific issues for discussion by Parliament. The above innovations mean that, once a year, Parliament may choose *any* current issue for further debate.

Individual ministries may have felt uneasy about the influence which these processes gave the Ministry of Environment to introduce changes in sector ministry policies (such as fisheries). During the preparation of Report No 46, there was strong opposition in some ministries to some of the ideas and proposals from the Brundtland Commission report, *Our Common Future* (WCED, 1987), being developed into national policies. As a result, there was a decision to lift the governmental steering committee up to the ministerial level. Eight ministers (led by the Minister of Environment) constituted the core group within the Cabinet. An outline of Report No 46 was considered by Cabinet first. It was decided to include a general part presenting major environmental challenges and global poverty problems, followed by sections on general policy and a chapter for each main sector, plus chapters on development aid. Each chapter outlined the situation, set out aims and presented new measures to meet challenges.

As usual in such processes in Norway, two substantial drafts of the report were prepared. These contained a number of proposals which were subsequently excluded following considerable internal political discussions between ministries. But the tensions ensured that all ministries gave the discussions serious attention. Some of the ideas in Report No 46 have taken several years to become adopted, for example, a change in the law to make pollution regulations apply to roads was due to have been considered by the Cabinet in April 1995. Report No 46 had been negotiated between ministries and there was political agreement on it, and the civil service had been intimately involved and had backed it. Consequently, the change of government in early 1989 did not derail the process.

The process of developing Report No 46 was charged with conflict between ministries and characterized by difficulties. Where civil servants could not resolve issues, they were passed to politicians to deal with. It was thus negotiated across government, approved by Cabinet and endorsed in Parliament. When the report was introduced in Parliament in April 1989, comments were overwhelmingly in favour. The recorded comments constitute a document as thick as the report itself. However, the sector ministries have made it clear that they would not wish to see another such report prepared — probably fearing loss of policy-making power.

(b) **The Bergen Conference and UNCED** Following submission of Report No 46, in 1990, Norway organized the Bergen conference as a regional follow-up to the work of the Brundtland Commission. A permanent committee on the environment was established, but was subsequently abandoned after a change of government. A representative national committee for the Bergen process was also established, comprising civil servants, labour organizations and NGOs. It held 13 meetings (approximately fortnightly) and continued to meet in preparation for and after UNCED as the National Committee for International Environmental Issues.

This conference was not accompanied by a new national policy report or by policy reform work, and consequently was not anchored in some way within ministries. But, after the conference, a list of measures that Norway should implement was drawn up and discussed by the Cabinet. Each ministry prepared a short statement outlining its perceived tasks. The conference secretariat in the Ministry of Environment reviewed the proposals in close cooperation with the department responsible for the issue concerned. Outstanding issues were resolved at meetings between the secretariat and the permanent secretary of each individual ministry concerned.

As a result of the work to prepare Report No 46 and the Bergen process, new policy guidelines were agreed, for example, for transport planning, coordination of land use planning and energy planning, and regulations on EIA have been implemented. After each of the UNCED preparatory committee meetings (PrepComs), a comprehensive report was prepared outlining all the positions taken, and giving the texts of all major

Norwegian interventions. This was circulated to the committee and all those in the civil service with an interest. The Norwegian position at UNCED was prepared with no political disagreement. An UNCED preparatory secretariat was also established, led by Mrs Eldrid Nordbø, former Secretary General of the Bergen conference.

(c) **Report No 13:1992–3** After UNCED, following Cabinet discussions, the Prime Minister issued instructions to prepare a report for Parliament on the implications for Norway of Rio agreements. The Norwegian UNCED preparatory secretariat led the process to prepare the Report to the Storting No 13:1992–3 'on the UN Conference on Environment and Development, Rio de Janeiro' (NorMoE, 1992b).

The report was prepared by a small editing team comprising the Directors-General of the International Department within the Prime Minister's Office, the Resource Department of the Ministry of Foreign Affairs, the then Ministry of Development Cooperation, and the Director-General of the Ministry of Finance (a former Ambassador to the OECD). With Mrs Eldrid Nordbø in charge of the process and Paul Hofseth as coordinator, it was completed within a six month period (July/August–December 1992).

The team interpreted Agenda 21, outlined what Norway had done to date and identified what remained to be achieved. It drew on all the ministries, mainly via very senior civil servants within the key ones. Ministries and individuals were advised on what kinds of material and approach were required for the report. If the team was not satisfied, submissions were returned for further elaboration or edited/written by the team itself.

While the aim of Report No 46 was to develop new policies for the environment, Report No 13 was to review existing policies in the light of the relatively undemanding texts from UNCED.

After UNCED, a Commission for Sustainable Development, chaired by the Prime Minister, was also established. Members include the chairpersons of the Confederation of Industry, the Labour Organization and the Nature Conservancy, and the Ministers of Finance, Foreign Affairs, Industry and the Environment (vice chair). The Ministry of Environment provides a secretariat. Initially the commission held some seven meetings, but has not been active for over a year. This commission has been a forum for exchange of views on important environmental matters before policy is finalized. It has discussed climate policy as well as the employment effects of environmental measures.

Prime Motivation and Getting Going

Report No 46 was initiated by politicians, primarily to elicit a response in Norway to the report of the Brundtland Commission (WCED, 1987). Prime Minister Brundtland herself wanted to establish a debate in public in Norway on sustainable development and to get a sustainable development strategy process going (unfortunately, the process met with considerable resistance and therefore may not have been what the Cabinet had intended). In practice, the focus of Report No 46 was mainly environmental, but the aim also was to include development aid and to use sectoral policies to achieve the objectives of the Brundtland Commission.

By comparison, Report No 13 was strongly inspired by civil servants from the most relevant ministries who had been engaged in the UNCED process. A measure of the relative difficulty is that Report No 46 took two years to prepare, while Report No 13 was completed in less than six months.

Terms of Reference

As with all reports to Parliament, both were initiated by decisions in Cabinet. For Report No 46, the State Secretary in the Ministry of Environment set out an initial operating mandate, which referred to the report of the Brundtland Commission and established an interministerial working group at State Secretary level. For Report No 13, the mandate reflected the need to report back to Parliament on the results from UNCED.

Participation

For Report No 46, the report of the Brundtland Commission was distributed to all local authorities and municipalities and to all NGOs (including environment groups, labour organizations and research institutions) for comment. The feedback was written up and published in summary form in Part 2 of the report (not available in English). It was also presented to the Committee of Ministry Secretaries of State but had little direct influence on the structure and content of the report. It was taken as reinforcement that the ministry had covered the right issues.

Before the Bergen conference, the Campaign for Our Common Future (*Felleskampanjen*) was established as an umbrella organization for all Norwegian organizations with environmental and sustainable development activities. Its membership comprised more than 100 such groups, ranging from farmers' associations to youth and church organizations, to the environment movement and women's groups. It coordinated participation in international events and received funding from several ministries. After Bergen, it was decided to extend its funding so that it could perform the same functions in preparing for UNCED. *Felleskampanjen* has now been succeeded by Forum — the campaign for development and environment.

In preparing for UNCED, a stakeholder committee — established originally for the Bergen conference — discussed issues with the civil servants preparing the Norwegian positions. An important point is that, until Cabinet decides to present policy papers to Parliament, they are secret and not open to public discussion, although some key ministries are usually deeply involved in the preparation and each are sent a copy 14 days before presentation. An innovative approach to parliamentary and public involvement is discussed above in section (a) on initiation and time perspectives.

NGOs were critical of Report No 46, seeing it as too growth-oriented, and as not tackling the fundamental difficulties of protecting the environment while aiming for growth. By comparison, Parliament was highly positive and most groupings declared that they wanted to strengthen the aims of the report. Again, this was reflected in the public debate.

Key 'Assistance' Factors, Problems and Issues

A number of factors helped in the preparation of both Report No 46 and Report No 13:

- public pressure on the environment;

- the Prime Minister's commitment to initiate a Norwegian response to the issue of sustainable development; and
- several comprehensive 'precursor pieces' of a green plan that had been written by Ministry of Environment staff in the period 1985–8 but had not been presented to Cabinet as a cross-sectoral plan. However, they were used internally within the ministry as a basis for thinking through environmental issues and organizing work.

There has been no centrally-organized systematic follow-up to Report No 46, not even in the Ministry of Environment itself. Plans were drawn up for action to be taken by each division within the MoE, but only some items have 'percolated through' to being implemented. Other ministries have not taken particularly strong action, except for the Ministry of Foreign Affairs (which took a leading role in preparations for and follow-up of UNCED) and the Ministry of Agriculture (which has promoted ecological farms, brought about changes in ordinary farming and reduced pollution considerably).

Climate change was a difficult issue, particularly the questions of how to stabilize CO_2 emissions at 1989 levels by 2000 and the establishment of a special climate fund (NorMoE, 1994b). Much effort was also placed on using economic instruments and this was crucial in securing support from the Ministry of Finance.

Links to National Planning and Decision-Making

Report No 46 was directly linked to the budget process, particularly in the innovation of the 'green budget' process. However, at its inception, it was poorly linked to economic planning, which is dealt with in the long-term programme. Report No 46 subsequently adapted itself to the long-term programme being prepared at the time and worked around what was being included in it, not the other way round.

Regional, Provincial and Local Strategies

Report No 46 takes account of existing Norwegian commitments under various declarations, such as those pertaining to the North Sea and to acid rain (NorMoE, 1992a). It also discusses issues concerning biodiversity and climate. Report No 13 discusses Agenda 21, the forest principles and the treaties on biodiversity and climate.

Each local community has been asked to make its own environment plan (see Box 16.1). Report No 46 aimed to stimulate county environmental and natural resource plans. Part 2 of the report contains a separate chapter on how local processes should proceed. An environmental plan has been prepared for Svalbard (Spitzbergen) by the Ministry of Environment.

Ecological Footprints and Transboundary Issues

Report No 46 discusses environment and trade issues, while the concept of 'ecological footprints' is dealt with implicitly in Report No 13 in its discussion of production and consumption. In January 1994, a sustainable consumption symposium was organized in Oslo (NorMoE, 1994a) and the ideas discussed were submitted to the Commission for Sustainable Development (CSD) at its May 1994 meeting. Subsequently, in February 1995, under a mandate from the CSD, a conference on sustainable production and consumption (the Oslo Ministerial Round Table) was held to prepare elements for an international work programme on these issues (NorMoE, 1995).

Cross-boundary issues, particularly acid rain and toxic chemicals, are a major point of focus for both Reports No 46 and 13.

Greening Lifestyles and Mainstreams

Neither Report No 46 nor Report No 13 have led to any extra greening of the political, business or consumer mainstreams. Norway was already relatively 'green'. The 'environmental groundwork' happened years ago, particularly during the 1960s and 1970s when there was intense national debate on green issues, and a Green Party was launched in Norway in 1970. Other parties then took up the green agenda. But Report No 46 led to institutional changes. It was important for Cabinet to see the environmental policies set out in Report No 46 in their totality and Parliament has used it actively.

Chapter 17 | POLAND

Izabella Koziell

The main strategy process in Poland is the National Environmental Policy of 1990, which was prepared by the Ministry of Environmental Protection, Natural Resources and Forestry (PolMEP, 1990) and was accepted by the Polish Parliament in 1991. The same ministry has also prepared an Implementation Plan to Year 2000 (PolMEP, 1994a). Other related activities include the report to the Commission for Sustainable Development on Poland's implementation of Agenda 21; and the Regional Environmental Action Programme of Central and Eastern Europe (REC, 1994).

National Environmental Policy

Initiation and Time Perspective

The main National Environmental Policy (NEP) document was drafted in 1990 and approved by the Council of Ministers towards the end of that year (PolMEP, 1990). It was accepted by Parliament in May 1991. The NEP commits Poland to sustainable development, which is stated as being the 'foundation of national environmental policy'. It sets out short-term (three to four years), medium-term (ten years, ie until the year 2000) and long-term (20–5 years, ie until 2020) priorities for environmental protection. In the initial stages, progress with the NEP was rapid and its formulation preceded even UNCED, but since the change in government in 1991/2, progress has diminished considerably.

The Implementation Plan to Year 2000, formulated by the Ministry of Environmental Protection, Natural Resources and Forestry in response to the NEP (PolMEP, 1994a), provides details of what goals need to be set and actions taken in order to implement the NEP's medium-term priorities. Other ministries have not responded actively to the NEP and key national policies (economic, energy, transport) have not yet shown any commitment to integrating environmental concerns into their policies, plans and projects.

Prime Motivation

The development of the NEP was the result of a series of events connected with the political situation in Poland. One of the initiating forces behind it was a growing awareness of the severe environmental damage suffered during the communist period and the need for radical policy reform. Repeated inaction and opposition by the communist regime led to the emergence of a strong environmental movement in Poland. The emergence of the Solidarity union in 1980 reinforced the environmental movement and provided a forum for public expressions of concern about the environment. The environmental movement grew in force and sought to inform the public of the linkages between environmental destruction in Poland, communism and central planning. The environmental movement supports the notion that government should integrate sustainable development concerns into sectoral policies and managed to have considerable influence during the 1989 round-table negotiations between the opposition and the communist regime. which led to the formation of the subgroup on the environment (discussed in section below on management and organization).

Poland's hope of entry into the European Union has also been a prime motivating factor in ensuring that the country meets EU environmental standards and legal norms as soon as possible. To achieve this goal, Poland will need careful strategic planning and considerable investment to comply fully with EU standards, regulations and procedures. Some standards will obviously need upgrading, while others (such as those of ambient air quality) are stricter than the EU's and may need to be downgraded.

Focus

The NEP departs from the previous narrow and centrally-controlled interpretation of environmental protection towards the broader goals of achieving sustainable development. It aims to achieve sustainable improvements in the state of the environment and to create conditions necessary for sustainable social and economic development. The basic principles of Polish sustainable development policy outlined in the NEP focus on:

- the control of pollution at source;
- the principle of being law abiding through reconstructing the legal system and strengthening its enforcement;
- the principle of a common good through the establishment of legal and institutional mechanisms to be enacted by citizens, social groups and NGOs;
- the application of the economization principle through using market mechanisms and the strict implementation of the polluter pays principle;
- regionalization; and
- the common resolution of European and global issues of environmental destruction; staging long-term plans through prioritizing activities.

The NEP also recognizes the importance of focusing on the reconstruction of the particular sectors of the economy that pose serious threats to the environment, i e energy, industry, transportation, mining, agriculture and forestry. It recognizes the need for a system of authority and responsibility in central and local government and outlines three different time frames of priorities for environmental protection: near-term (three to four years), medium-term (up to the year 2000) and long-term (until 2020). The environmental policy tools that will ensure effective implementation of the NEP include legal and administrative measures, economic instruments, inspection and monitoring systems, environmental education and research.

The Implementation Plan to Year 2000 was developed in response to the NEP and sets out details of what investment and non-investment tasks are necessary to achieve full implementation of the medium-term priorities articulated in the NEP. It was formulated on the basis of responses to a questionnaire sent to local government offices throughout Poland. It specifies the estimated costs of required investments, states what administrative and legislative changes are necessary and sets targets by which to measure progress. The following key issues are addressed:

- air protection against pollution from stationary sources;
- air protection against pollution and noise control from transport;
- protection of water bodies;
- nature protection;
- management of particulate/dust emissions;
- water resource management;
- forest management;
- mineral resource management;
- land management;
- noise and vibration control;
- control of non-ionizing electromagnetic radiation;
- management and control of environmental emergencies; and
- establishment of an integrated approach to environmental management.

Responsibility for implementation of some of the stipulated tasks relate to other key sectors, such as energy, transport, housing, and agriculture. These sectors have not shown much interest in integrating environmental concerns into their policies, so potential success could therefore be limited.

The Ministry of Environmental Protection, Natural Resources and Forestry is now required to produce a triennial report on the 'analysis and implementation of the national environmental policy'. The most recent one (1991–3) reported on the progress of implementing the short-term priorities stated in the NEP (PolMEP, 1994b).

Management and Organization

In 1989, during Poland's transition from central planning to a parliamentary democracy and market economy, a 'round table' was established which aimed to provide a forum for serious discussion between the opposition and communist regime. Increasing concern over the state of the environment in Poland and pressure from outside groups led to the establishment of a subgroup on the environment which would provide a forum for discussion of the future direction of environmental protection in Poland.

The round table's subgroup on the environment brought together some of Poland's most distinguished environmental experts and the group consisted of key individuals from academic institutions, NGOs and environmental movements in Poland. It was viewed as the forum that would have the power and ability to commit Poland to the path of sustainable development, through agreeing to a series of reforms and new policies. After two months of intensive discussions, 27 recommendations were put forward, which laid the foundations for Poland's NEP. However, the initial enthusiasm was not carried forward and those given responsibility for implementing the proposed recommendations were hampered by administrative bureaucracy and political instability (President Walesa's *wojna na górze*). However, four radical moves were made in 1991/2 which included:

(1) pollution fees were elevated to their historically highest levels. Subsequently, after a series of revaluations, they declined in real terms; but, after 1994, two further revaluations have resulted in fees regaining their level in January 1992;
(2) the State Environmental Inspection Act (1991) established environmental inspectors (or 'police') independent of regional administrators — something that the current coalition has tried to reverse;
(3) the Forestry Act (1991) introduced very favourable taxation of forestry in order to promote sustainable uses of forests;
(4) the Nature Protection Act (1991) established unusually broad competencies for national park directors — on national park territory they have the authority of a provincial administrator (*voivoda*).

Towards the end of 1989, concern over environmental policy inaction was growing. The Parliamentary Committee for Environmental Protection, Natural Resources and Forestry, which had existed since the 1970s, and the Ministry of Environment, Natural Resources and Forestry began work on proposed policy and legislative changes, but progress was frustratingly slow. Some observers believe that new proposals emanating from the ministry stagnated within the committee (which kept changing its political alignment). Others believe that the ministry was responsible for the lack of action. It is said that the committee can only take credit for the Nature Protection Act (1991), while the ministry achieved all the other changes mentioned in (1)–(3) above.

This process did, however, lead to the emergence of the NEP and it raised environmental awareness within Parliament, which helped in achieving parliamentary approval of the NEP document. However, the lack of progress towards finalizing an Environmental Protection Act (new versions are still being discussed) has more to do with the academic community than with the political circles.

The Ministry of Environmental Protection, Natural Resource and Forestry is now responsible for development and implementation of the NEP. Several other ministries also have important roles to play in the NEP process. These include the:

- Ministry of Spatial Planning and Construction (municipal water supply, sewage collection and land use planning);
- Ministry of Health and Social Welfare (environmental hygiene monitoring, food and drug monitoring);
- Ministry of Agriculture and Food Economy (pesticide control);
- Ministry of Labour and Social Policy (occupational safety);
- Ministry of Industry and Trade (mining, energy production and manufacturing); and the
- Ministry of Transport and Maritime Economy (regulations on car exhaust gases); and its scientific research committee (research and development issues).

At a national policy level, other institutions involved in an advisory role include the Environment Council, established by the President of the Republic in 1993; and the

Council for Sustainable Development, established by Prime Ministerial decree in October 1994 (this body has now become an important means of integrating government policy). Other bodies established by the Minister of Environmental Protection include the State Council for Environmental Protection, the State Council for Nature Conservation and the river basin councils.

Implementation of the NEP at the provincial level will be carried out by the 49 *voivodship* (provincial) authorities, except for river basin management, which is performed by regional water management authorities. Each *voivodship* has a department of environmental protection, which manages and coordinates environmental activities in the province and develops local policies consistent with the NEP. At a more local level, the *gmina* (a local administrative body typically comprising a dozen or so parishes) takes on certain environmental responsibilities, which include the municipal water supply, waste water treatment and waste management, local level nature conservation, land-use planning, and management. Monitoring environmental regulations and issuing local permits for certain investment projects are effected by the *voivodship*.

The State Inspectorate of Environment, which was established in 1980 but gained more power under the 1991 law on the inspectorate, enforces and monitors environmental laws. It is made up of both state and *voivodship* inspectorates and has a network of laboratories.

Participation

The public did not participate in the development of either the NEP or the Implementation Plan to Year 2000. The latter was formulated by a panel of Poland's leading environmental experts and, during the course of this process, some efforts were made to communicate with the *voivoidas* through a questionnaire. However, the questionnaires were aimed at collecting factual information and not at providing options for national environmental policy.

Recent changes in the physical planning system (January 1995) have upgraded the role of local governments, empowering them to take more decisions on the environment, but the Ministry of Environmental Protection, Natural Resources and Forestry has not yet developed suitable legislative and institutional systems to facilitate public participation. However, the minister has agreed to include representatives of NGOs and local governments in the State Council for Environmental Protection, the State Council for Nature Conservation and in the river basin councils, where local government representatives have also been brought into the councils supervising the *voivodships*' funds.

The relevant authorities have been slow to change the habits they picked up under the communists, when information was highly controlled and dissemination restricted. This has caused a delay in, and limited the extent of, the dissemination of environmental information to the public. However, some information is available in environmental statistic yearbooks.

Links to NGOs and the Public

The NGO environmental movement had a decisive influence during the round-table negotiations in 1989. But since then environmental NGOs appear to have diminished in size and importance and seem unable to present a united front. Recently, NGO involvement at the national policy level has been very limited, with most focusing on local level issues. In 1992, the Ministry of Environmental Protection, Natural Resources and Forestry began to arrange meetings between the NGOs and the minister, but since 1993/4 they have not taken place, though the ministry recognizes their importance and intends to revive them.

However, there are a large number of small groups, comprised mainly of young people, that are fairly active 'watchdogs' at a local level. Larger organizations include the Polish Ecological Club and the Nature Conservation League. A small National Environmental Education Centre functioned during the period 1993–4 and created a network of 20 regional environmental education centres (most of which are typical NGOs) which still exist.

There are plans to set up an 'environmental education' group to provide a forum for discussing an environmental education strategy and options for NGO cooperation. NGOs have been particularly effective in environmental education by organizing seminars, conferences and lectures for various target groups, as well as various actions for the environment, such as Earth Day.

There are no MPs representing a Green Party. However, some (for example Mr Radoslaw Gawlik) are well-known environmental activists.

The NEP has been classified by some NGOs as a 'paper policy' advocating traditional environmental protection. NGOs feel that the ministry is not at all interested in involving NGOs in the process. The Implementation Plan to Year 2000 was not released to NGOs during the review process and is seen very much as an internal document. However, NGOs can now receive copies of proposed environmental legislation before its submission to Parliament. Nonetheless, there are plans to involve NGOs in the process of formulating the framework for environmental law.

NGOs are represented on the Commission for Environmental Impact Assessment, which is responsible for evaluating and approving environmental impact statements; and NGO participation is therefore institutionally guaranteed.

Key Problems and Issues

Enthusiasm generated for the NEP in the post-communist era has faded as the realities of democracy, the market economy and consumerism set in. There is huge pressure to try and 'catch up' with the living standards of Western countries and to become members of the European Union while at the same time safeguarding the environment. The NEP makes an attempt to tackle this issue, and clearly identifies the responsibilities of the sectors. But there is still little evidence of the integration of environmental concerns into other sectoral policies and plans. In cases where policies contain the rhetoric of sustainable development, there is no indication of how such policies will be implemented. The Ministry of Environmental Protection, Natural Resources and Forestry is the only ministry to have produced an implementation plan as a result of the NEP — even though other ministries are key players in the implementation of the NEP. It has not been possible to integrate policies, as was originally planned.

The present socialist government is moving back towards centralization and sectoral thinking typical of socialist economies. This is illustrated in the long-overdue sectoral policies (for example, energy, transport, agriculture, privatization, tourism), which seem to fail to acknowledge the need to integrate environmental concerns. The government is not addressing the issue of how to maintain current levels of economic progress while at the same time protecting the environment. Generally, it is more interested in short-term economic gains.

The continual Cabinet changes and reorganization within the Ministry of Environmental Protection, Natural Resources and Forestry, as a result of the unstable political situation, have prevented the development of secure policies. Also, with so many staff changes, there has been a loss of institutional memory. In 1990/1, the ministry had considerable influence, but it is now one of the weaker ones and, having lost its political weight, can no longer exert much political pressure on other sectors.

The early success in formulation of the NEP in Poland meant that it served as a model for many other central and eastern European countries. The economic instruments and legislative mechanisms developed as part of the NEP process are considered to be the most advanced in the whole of central and eastern Europe, but there has been too little use and development of other tools such as EIA and voluntary agreements.

Regional Strategies

The 'Green Lungs of Poland' — a government initiative started in 1991 — aims to combine nature protection with economic development in northeastern Poland (NFEP, 1991). This is an area very rich in biodiversity and the strategy was developed with participation of the *voivodships*. It has already resulted in abandonment of plans to open a mine. More recently, it has been extended as the 'Green Lungs of Europe' initiative encompassing a much larger area — Belarus, Estonia, Latvia, Lithuania, Poland, Russia and the Ukraine, and the 'Green Lungs of Europe' international agreement was signed in Warsaw in February 1993 (ISD, 1993).

A strategy is being planned to preserve the lower reaches of the Vistula River, which contains the only nesting grounds in the whole of Europe for particular species of birds. Assistance is being sought from the IUCN.

Local Strategies

While there were already several local strategy initiatives operating in Poland, including the Radom Project for Sustainable Communities (discussed on p59), in 1995 each *gmina* became obliged to develop a local sustainable development strategy (Local Agenda 21) within the next five years.

Convention-Related Strategies and Transboundary Issues

Poland has adopted over 40 international conventions related to transfrontier air pollution, pollution of the Baltic Sea, transfrontier movement of hazardous waste, protection of the ozone layer, climate change and nature protection. At a global level, Poland has ratified the Montreal Protocol and the UN Framework Convention on Climate Change, and is currently ratifying the Biodiversity Convention. Progress has been made in the implementation of several obligations, particularly to eliminate ozone-depleting substances and to reduce SO_2 and greenhouse gas emissions. The ratification and implementation of these conventions have serious financial and administrative implications for the country.

Poland has signed various agreements with neighbouring countries concerning water management, water pollution, nature protection and air pollution. In the case of water and air pollution, Poland acts as a pollutee rather than a polluter.

Parliamentary Process and Wider Debate

The NEP was discussed in Parliament and the realization of the Implementation Plan to Year 2000 is still under discussion. However, the question of integrating environmental concerns into other key sectoral policies tends to be ignored.

The NEP process may have raised some environmental awareness in government but this has not been translated into action by the other key sectors. There has been very little discussion on the 'greening' of consumer and business mainstreams. But a cleaner production movement has been developing within the business community. Businessmen are invited to participate in the advisory councils to the Minister of Environment and can provide comments on draft environmental regulations.

Press Coverage

Some radio and TV programmes are environment-related and occasional articles appear in the daily newspapers and magazines. Reporting on the environment is not, however, an urgent priority. There is a lack of 'environmental reporting' skills and insufficient understanding among journalists of the issues at stake, and the roles and responsibilities of the various institutions. This sometimes results in incorrect reporting. There are several periodicals, produced by NGOs, which deal specifically with environmental issues. The ministry has organized press conferences to inform the public on the latest developments in environmental protection. At the same time, the media also promote western-style high consumption models which are frequently contrary to the principles of sustainable development.

Chapter 18 SWEDEN

Several initiatives contain elements that are considered to comprise the equivalent of a green planning or sustainable development strategy process in Sweden. Among these are:

- *UNCED*, Swedish Government Bill 1993/4:111 (submitted to Parliament on 9 December 1993, adopted in late April 1994) (SwedMoE, 1994).
- *The Environment: Our Common Responsibility* (draft 95–03–08), (Government Communication 1994/5:120 presented to Parliament on 20 December 1994) (SwedMoE, 1995) — an elaboration of the 1994 bill by the new Swedish government outlining the focus and strategy of its environmental policy, and a state of the environment report.
- Every year, the Ministry of the Environment and Natural Resources prepares a state of the environment report for discussion by Parliament.
- *An Environmentally Adapted Society: The Action Programme of the Environmental Protection Agency: Enviro '93*, prepared by the Swedish Environmental Protection Agency (SwedEPA, 1993). This 'strategy' also raises issues associated with the UNCED follow-up (see section below on an environmentally adapted society: the action programme of the Environmental Protection Agency: Enviro '93).

Towards Sustainable Development in Sweden: Government Bill 1993/4:111

Initiation and Time Perspective

In August 1992, following UNCED, a representative national meeting was convened to consider what follow-up action should be taken in Sweden. Subsequently, in Spring 1993, eight regional meetings were organized to convey the messages of UNCED and to discuss their implications with all interested actors.

Each year, the Ministry of the Environment and Natural Resources (MoE) prepares a report for discussion by Parliament, which includes elements of environmental policy as well as a state of the environment assessment. As a follow-up to UNCED, the MoE decided that the regular report for 1993 should be prepared as a government bill. In October 1992, a White Paper was presented to Parliament outlining what had happened at UNCED and what follow-up action the government intended to take.

Soon after UNCED, Agenda 21 and the Rio conventions were translated into Swedish. These documents were circulated to about 300 organizations in the country, including government departments, local authorities, NGOs and individuals, for review and comment. In Spring 1993, on the basis of a joint letter from the MoE and the Swedish Association of Local Authorities, all local authorities were asked to consider how UNCED agreements tallied with the present situation and policies in the country and to recommend what changes were needed. A summary of Agenda 21 was prepared and 25,000 copies distributed nationally. As a result, 'Agenda 21 is now a household word in Sweden'. The MoE still receives 10 to 15 calls per day requesting copies of the summary. About 250 organizations made written responses proposing actions/changes. In discussing these responses, the bill (p11) notes that:

> *The general consensus is that the future environmental objectives expressed in the Rio Declaration on Environment and Development and Agenda 21 are as essential as they are correct. The action programme must, however, be made more precise in terms of goals and action plans locally, regionally and globally. The vast majority of statements express a strong will to act, but also stress the lack of resources.*

In a separate communication, all 285 local authorities were invited to embark on preparing Local Agenda 21s; some 220 are now engaged in developing these. They represent 'more of a process than plans'.

The MoE analysed the public responses and prepared a draft outline of a bill. This was circulated to ministries for review and comment. Many rounds of negotiation followed. The bill covers the period up to 1997, as well as including aspects of a more long-term nature (see Box 18.1). It focuses mainly on domestic issues, but it also contains a chapter on Sweden's global cooperation. Special attention is given to development cooperation, environment and trade, and conventions and treaties. The bill does not focus specifically on 'ecological footprints' but implicitly recognizes the issue in discussing changes needed in production and consumption patterns. Transboundary issues, particularly acid rain, are also dealt with, but these were serious issues well before UNCED.

The opposition Social Democrats had wanted a national Commission for Sustainable Development to be established, but the then government did not favour this approach. Following elections in 1994, the new government introduced such a commission (chaired by the Minister of Environment) to take up sustainable development issues with all actors and stakeholders and to develop a basis for a Swedish report to the forthcoming 1997 special session of the UN General Assembly. Follow-up to the 1994 government bill will also be coordinated through the commission, which will review implementation of UNCED decisions.

Prime Motivation and Getting Going

The main imperative for initiating the bill was a desire by the MoE to make progress on translating the agreements reached at UNCED into policies and action within Sweden. The focus was on the overall agenda of sustainable development. This was viewed as a major leap forward from the early 1990s when 'green attention' was given only to the environment. There has been a long tradition of environmental policy-making in Sweden over the last 30 years. The first environmental protection legislation was introduced in 1968 when the Environmental Protection Agency was established. There has

Box 18.1 *SUMMARY OF SWEDISH GOVERNMENT BILL 1993/4:111*

The official summary of the bill states:

One of the Government's priority areas is to work towards long-term sustainable development in Sweden within the framework of the carrying capacity of the environment.

The Bill lays down general guidelines for development in Sweden within the various problem areas and sectors of society. Together, they form a national strategy for sustainable development and thus constitute a Swedish Agenda 21 based on environmental goals and priorities already adopted by the Riksdag [Parliament], inter alia in terms of the development of an ecocycle society, climate questions, biological diversity and forestry policies and matters currently at the preparatory stage, such as an environment code, and questions concerning chemicals and the environmental debt. A report on the state of the environment — the government's annual environmental report to the Riksdag — is appended to this bill.

The government intends to take action to ensure that environmental impact assessments are performed as a general rule either before or in conjunction with strategic decisions either by the government or by state bodies. Such bodies should consider the environmental implications before they issue regulations or guidelines. Instructions, statutes, etc, will then be reviewed for assessment of the necessary environmental adaptation.

The government also intends to give state bodies the possibility of assuming special environmental responsibility in their activities. Such responsibility will be reviewed during the course of future environmental audits.

The government intends to support local authorities in their work on drawing up local action plans for sustainable development. In order to achieve sustainable development, better management of resources will be required, along with changes in patterns of production, consumption and lifestyles. The bill discusses the way in which local authorities can contribute to such development, in conjunction with a Local Agenda 21. In the 1994 draft budget, the government intends to consider a proposal submitted by the enquiry on the work of local authorities on a good living environment concerning the promotion of work on Local Agenda 21 programmes in local authorities and NGOs.

The government intends to instruct the Environmental Advisory Council to develop the follow-up of the Rio decisions in Sweden.

This bill also discusses the Swedish strategy for future work on questions relating to the global environment and development. Goals, priorities and measures are presented, as well as relevant organizational issues and questions concerning resources. A special working group is to be appointed to deal with environmental issues and development cooperation issues in the light of the Rio decisions.

Source: SwedMoE, 1994.

been a tremendous movement among businesses to become competitive on environmental grounds. But there has also been a great shift over the last few years towards a common national understanding of the concept of sustainable development and sound environmental management. The bill process was therefore aided by strongly supportive public pressure and political understanding. It was envisaged that the 'clout of Rio' could be used to make in-roads on finance, for example by amending the taxation system. The Ministry of Finance agreed to set up a committee to investigate ecological tax reform. It was also persuaded to state that Sweden's 'environmental debt' (i e what Sweden owes nature, *c.* 800 billion Kronor) should not increase. This statement has not been reconfirmed by the new government, although it has not been negated either.

Organization and Management

Terms of reference for preparing the bill (the process and an outline content) were drawn up by civil servants in cooperation with political advisers in the MoE. A working group of individuals from the various departments in the MoE was established. Each individual took responsibility for the development of a particular chapter of the bill. A second working group was set up comprising representatives from each ministry to discuss emerging ideas and to negotiate issues and problems.

In all, the process took less than one year. The UNCED documents were circulated for comment to 300 organizations in February 1993. In mid June, a number of people were hired to collate the responses. The bill contains a summary of the responses received for the benefit of Parliament. Ideas on how to proceed were circulated to ministries in mid-September. The bill was concluded for printing on 23 December 1993. This process was similar to that used for preparing all major bills in Sweden. Some NGOs took the view that the bill represented 'too little to late' but, in general, the political and public response was favourable.

Key 'Assistance' Factors, Problems and Issues

The development of the bill was greatly helped by the fact that the Minister of the Environment at the time (1992/3) was also leader of one of the coalition partners in government. He had attended UNCED and was particularly committed to the bill. However, it was not easy to decide which areas to focus on in developing the bill, since Sweden had had an established environmental policy for many years. It was felt that there were relatively few aspects of Agenda 21 that were new issues in the country. A decision was taken to pursue a thematic approach, for example, dealing with issues such as the 'polluter pays' and 'precautionary' principles, and strategic environmental assessment. The bill states, for example, that:

> *It is important for environmental impact assessments to be systematically integrated into the political process from an early stage. Such assessments should be employed in conjunction with programmes, plans and policy work in the main sectors of society. The government considers that, as a general rule, environmental impact assessments are to be included in government bills and other proposals concerning general decisions of a strategic nature. With the procedures proposed here, analysis of the environmental impact of major strategic decisions will become standard practice.*

> *Authorities should take environmental impact into account before issuing regulations or general guidelines. The government intends to carry out a review of the rules, ie instructions and statutes, etc, in order to assess the need of environmental adaptation.*
>
> (SwedMoE, 1994, p26)

Other ministries were particularly concerned with 'safeguarding' their positions during negotiations on the bill. Many ideas on sustainable development being discussed during preparation of the bill were able to be introduced into other bills in other sectors being developed at the time. For example, the Ecocycle Bill (March/April 1994) adopted many arguments expressed in Agenda 21 and in bill 1993/4:111. The concept of the 'ecocycle society' is based on the principle that

> *whatever is extracted from the natural environment is to be used, recycled and disposed of without injury to the natural environment. This means that renewable resources which play a part in nature's own ecocycle should be used wherever possible. Toxic chemicals that are not readily degradable and toxic metals are not to be used.*
>
> (SweMoE, 1994, p8)

Conflict Resolution and Consensus

Most problems were negotiated during continuous discussions and bargaining. In the particular circumstances of a coalition government, the special coordination unit established at the Prime Minister's Office, and representing all four parties, was able to resolve some of the interministerial conflicts over elements of the bill. As with all bills negotiated for presentation to Parliament, consensus was reached across the various ministries.

Links to National Planning and Decision-Making

The bill provides general guidelines, but it is not an act with the force of legislation. Neither does it contain budgetary implications — perhaps a reason why it was possible to secure agreement and parliamentary approval. But it has influenced legislative changes and is frequently referenced. Many elements of the bill, and issues discussed in it, are being used in discussions and negotiations to develop Local Agenda 21s and in local planning in Sweden. The bill led to the 1994 Finance Bill mentioned above, in which 10 million Kronor were set aside for Local Agenda 21s.

An Environmentally Adapted Society: The Action Programme of the Environmental Protection Agency: Enviro '93

This action programme was prepared by the Swedish Environmental Protection Agency (SwedEPA, 1993). The English translation actually uses the title, *Strategy for Sustainable Development* (SwedEPA, 1994), although, in practice, it is an action programme that raises issues associated with the UNCED follow-up.

Initiation and Time Perspective

Enviro '93 was initiated in Autumn 1991 as an internal initiative of the Swedish Environmental Protection Agency (EPA). The document, submitted to the government in Summer 1993, assessed (with a 5–10-year perspective towards 2000) trends in the state of the environment and the environmental impact of different sectors of society, and proposed a wide range of measures (see Box 18.2).

Suggestions and data from Enviro '93 were also taken up and used in the preparation of the government bill (1993/4:111), *Towards Sustainable Development in Sweden* (see section above on towards sustainable development in Sweden: Government Bill 1993/4:111). Enviro '93 followed on from several sectoral action plans prepared in the 1980s, for example, for air, marine pollution, land use, biodiversity and nature conservation, and from the 1988 *Environment Policy for the 1990s.*

EPA is now working with stakeholders on preparing a set of parallel sector action plans, which have a vision to 2021.

Prime Motivation and Focus

Previous action plans had been media-related (focusing on air, land, sea, etc). It was felt that a shift was needed to focus on action needed to be taken within different sectors of society. The trend was also for environmental policy as a whole to move towards a sector-oriented approach, with each sector assuming its own responsibility. The aim of Enviro '93 was to show how individual sectors would need to respond in developing sectoral environmental action plans and programmes. Most of the goals, objectives and targets set by Swedish commitments under international and regional conventions and agreements are included in Enviro '93.

Organization and Management

An advisory group was established within EPA comprising heads of divisions. In addition, a steering committee was set up. Two separate sets of reports were developed: (a) background reports and (b) action plans.

(a) The following (green) background reports were produced (those with numbers beginning with 42 are translated into English):

Acidification: An Everlasting Problem, or Is There Hope? (SNV Report 4242).
Ground-level Ozone and Other Photochemical Oxidants in the Environment (SNV Report 4243).
Eutrophication of Soil, Fresh Water and the Sea (SNV Report 4244).
Metals and the Environment (SNV Report 4245).
Persistent Organic Pollutants and the Environment (SNV Report 4246).
Environmental Pollution and Health (SNV Report 4249).
Land Use and the Environment (SNV Report 4137).
Biodiversity (SNV Report 4138).

Box 18.2 SCOPE OF SWEDEN'S ENVIRO '93 PROGRAMME

According to its Foreword, Enviro '93:

- *describes the environmental situation in relation to the environmental objectives, and assesses the gravity of various problems;*
- *follows up international commitments and decisions adopted by the Swedish government and Parliament (the Riksdag), and*
- *proposes measures in different sectors of society, as well as in international cooperation.*

In some areas of environmental protection, such as waste management and chemicals control, we are in the process of implementing several very ambitious political decisions. In these cases, the emphasis of the action programme is on reporting on the different efforts being made to implement decisions that have already been adopted. In other cases, such as forestry policies and measures towards the establishment of a society based on the principles of ecological cycles, new objectives and guidelines for environmental work have just been passed by the Swedish government and the Riksdag. In these cases, the emphasis of the action programme is on discussing how working methods, etc can be developed to bring the general political objectives to realization.

It has been our intention to identify the main environmental challenges and the measures needed in various sectors of society. In some cases, we have made detailed proposals for measures, while in other cases we have made less progress in stating specifically what needs to be changed, and what the cost will be. The strategy offers a common platform for continued work to be done by the Environmental Protection Agency and other government agencies, county and local authorities and, of course ultimately, by every relevant sector of society.

Source: SwedEPA, 1993.

(b) A set of (yellow) action plans were prepared covering energy, traffic, industry (only this has been translated into English), water and sewage, agriculture, forestry and nature conservation.

Each of the reports in these two sets (except the action plan for agriculture, led by the Board of Agriculture) was prepared by a project group comprising EPA staff, representatives of scientific groups and individual experts. There was also an overall steering group for all the reports comprised of staff from EPA and representatives of county administrative boards and local government. Drafts of each issue report were distributed for comment to scientific organizations and sector agencies (forestry, agriculture). The background reports were consolidated to produce the environmental overview in Enviro '93, while the action plans were the basis for the second part of the document. Additional elements and suggestions were also included.

The scientific response was very positive with consensus on the problems identified in the document. Special articles were written for magazines, and seminars and conferences were organized. Briefings have also been provided for local communities developing Local Agenda 21s to enable them to use the information presented in Enviro '93.

Each of Sweden's 24 counties has been asked by the government to develop its own environmental policy plan. Through Enviro '93 follow-up, EPA is trying to stimulate communication between the county plans and Local Agenda 21 initiatives.

Terms of Reference

Terms of reference were by set by the EPA in a project plan. The Ministry of Environment then gave general approval. One of the main initiators in EPA became an Under-Secretary in the MoE and was able to continue to support the approach.

Participation

The preparatory phase was not open to the public, but was controlled by EPA (see section above on organization and management). The report project groups for forestry and agriculture included representatives appointed by the national agencies responsible for those sectors. The groups also included individual experts invited from counties and municipalities with which EPA had close working contacts. Other national agencies — such as those for road traffic, railways, aviation, and fishing — participated through meetings, contributing information and giving their views on drafts. There were no representatives of the private sector or of NGOs.

The (yellow) action plan for agriculture was prepared collaboratively by the Swedish Board of Agriculture and the EPA. All the others were led by EPA. The overall Enviro '93 document was prepared mainly by EPA staff, but individuals on the report project groups were able to review particular sections, and some commented on the whole document. The published document was circulated for comment to universities, regional and local government, industrial organizations, NGOs and other agencies.

Some 3000 copies were published. There was considerable interest from the press but relatively little national debate. NGOs responded positively but expressed a view that more needed to be done.

Key 'Assistance' Factors, Problems and Issues

The government had placed emphasis on the need for a sectoral approach to dealing with environmental issues, and had made sector agencies responsible for elaborating their own environmental programmes. The county administrations had also been charged to prepare environmental policy plans and were keen to receive Enviro '93 to assist them in their tasks. Major problems were the limited manpower available for the project, and the two-year time frame set by EPA which it then had to meet.

Key issues addressed in developing Enviro '93 included identifying as many environmental quality goals and objectives as possible, and setting costs to compare cost-effective ways of solving problems — there were too few economists available at the time.

Conflict Resolution and Consensus

Most difficult problems were negotiated, first, at meetings at the project report group level, then at higher level and, finally, between directors-general of organizations. However, Enviro '93 acknowledges that some of the proposals contained in the document are EPA's and were not agreed to by other agencies, and that it will be up to government to make decisions on these at the appropriate time.

Written responses suggest that the document was generally well received by the government, county administrations and other agencies. For example, Enviro '93 is much used by communities preparing Local Agenda 21s and by county administrations developing their environment policy plans, to set environmental targets.

Ecological Footprints and Transboundary Issues

There is some discussion of future consumption patterns, and trade and environment issues are emphasized. Trade agreements are discussed as a tool for environmental management. Important cross-boundary issues considered are acid rain, eutrophication and metal pollution (particularly mercury and cadmium from eastern Europe).

Parliamentary Process

Enviro '93 as a whole has not been subject to any governmental or parliamentary process. However, certain elements of the document have been taken forward through government bills (for example, Bill 1993/4:111; Biodiversity 94).

Greening Lifestyles and Mainstreams

Enviro '93 is one of many initiatives that have contributed to making Sweden 'green'. It has strengthened particular ideas and concepts. Meetings were held with political parties and leaders, and with parliamentarians, to explain its contents. Progress in greening industry was limited as much had already been achieved.

Ongoing Work

Enviro '93 is expected to remain valid for some years and there are no immediate plans to revise it. EPA is now working on elaborating a vision of an environmentally-adapted society in 2021 in a three-year programme (1994–7). This involves discussions with all stakeholders and sectors, using participatory approaches (for example, Delphi techniques).

Chapter 19 UNITED KINGDOM

The UK has developed a linked suite of strategies.

- This Common Inheritance: Britain's Environmental Strategy (HMSO, 1990);
- Sustainable Development: The UK Strategy (HMSO, 1994a);
- Climate Change: The UK Programme (HMSO, 1994b);
- Biodiversity: The UK Action Plan (HMSO, 1994c);
- Sustainable Forestry: The UK Programme (HMSO, 1994d); and
- Local Agenda 21 initiative (coordinated by the Local Government Management Board), 1992 and continuing.

The climate change programme, biodiversity action plan and sustainable forestry programme were prepared as 'parallel' documents to the UK sustainable development strategy and were published together with it. The strategy discusses and integrates the other programmes/plans in some detail.

Sustainable Development: The UK Strategy

Initiation and Time Perspective

In response to the Brundtland Commission's report (WCED, 1987), the government published a comprehensive UK strategy for the environment in a White Paper, *This Common Inheritance* (HMSO, 1990). It included many specific targets (for example, for CO_2) and objectives (such as to take precautionary action) for policies in different areas, and set out various commitments (for example, to cut the government's energy bill). Progress on these, together with new commitments, have been reported in annual update reports, initially released each Autumn (HMSO, 1991; 1992), and subsequently rescheduled for the Spring (HMSO, 1994e; 1995) following publication of *Sustainable Development: The UK Strategy* in January 1994. No actual update report was produced in 1993, as work on the UK strategy was in hand. A Cabinet Subcommittee on the Environment was established, which introduced a new rule that any policy proposals going before Cabinet should be accompanied by a statement of environmental impacts where these are significant. The DoE set up consultative committees with local government and the voluntary sector, and — acting jointly with the Department of Trade and Industry — with industry. 'Green ministers' were also 'invented':

> *In each government department, one of the ministerial team now has responsibility to make sure that its policies and management have environmental concerns built into them. Some, but not all, of these Green Ministers are full members of the Cabinet. They work within their own departments, and meet every few months to compare notes on progress and discuss common problems.*
>
> (Stevens, 1995)

At UNCED in 1992, the British Prime Minister, John Major, committed the UK to producing a sustainable development strategy by the end of 1993. A main motivation was that the UK should respond (and be seen to do so as an example to others) speedily to Agenda 21 and the other UNCED accords.

In September 1992, thought was then given within the Department of the Environment (DoE) on how to undertake such a strategy. While the 1990 environment strategy provided a substantial basis for a sustainable development strategy, it was considered to require more than 'just updating' to take account of UNCED (although some people argued for this). It was decided that a new strategy process was required and proposals in this regard were developed and discussed within government through bilateral discussions between the DoE and other departments, and at a collective interdepartmental meeting. The broad approach was subsequently approved by Cabinet and the Prime Minister. In the Spring of 1993, substantive discussions on the strategy approach took place within an official interdepartmental committee comprising grade II departmental secretaries and chaired by the Cabinet Office. The committee continued to consider the matter at several subsequent meetings. In the summer of 1993, the Cabinet Subcommittee on the Environment discussed and approved a 'consultation paper' covering the scope and general structure of the strategy, and later in the year endorsed the draft strategy. It was then presented to Parliament collectively by 16 Cabinet ministers with responsibilities for some aspect of sustainable development. Throughout the process, the Prime Minister's Policy Unit was actively involved in debates and correspondence.

The strategy has an overall perspective to 2012 — 20 years after Rio. However, in some cases, a longer time frame is involved, for example forecasts for greenhouse gases are made up to 2020.

The fourth annual update report (HMSO, 1994e) covered an 18-month period (October 1992 to May 1994) and was a document which represented a 'confluence of two streams' — reporting on progress of both the 1990 environment strategy and the 1994 sustainable development strategy.

Focus

The main focus of the UK strategy is sustainable development — from the government's perspective. It set out to examine 'not only environmental protection, but the essential relationship between this and economic activity' (Stevens, 1995). While this linkage was contained in the 1990 environmental strategy (HMSO, 1990), the sustainable development strategy set out to make it more explicit. The government took the view that social issues need not be a prominent theme, and that they would be dealt with as they 'ran in and out' of the issues of primary concern linked to the above mentioned focus. Consultations on the focus of the strategy supported this approach —

most groups were mainly concerned with the balance between wealth and environmental protection.

The strategy sets out various principles for sustainable development covering the need for economic development, caring for the environment, specific principles such as the precautionary principle, and mechanisms for taking the right decisions. There are three main sections. The first discusses the state of the environment and issues related to the various media (for example, air quality, the sea and soil) — looking 20 years ahead or more. The second considers economic development and sustainability, discussing likely impacts on the environment from different economic sectors (for example, forestry, minerals extraction and transport). The final section discusses different types of policy responses (putting sustainability into practice in the context of, for example, the international perspective, voluntary organizations, the land use planning system, and energy efficiency).

Each chapter starts with a standard checklist of points: a sustainable framework; trends; problems and opportunities; current responses; and the way forward. The UK strategy is strong on analysis, clearly setting out principles. Stevens (1995) points out that 'it declares that the best way to harmonize economic development and environmental protection is to bring environmental costs and benefits into the heart of economic decisions — both in government and in the private sector. And it confirmed the British government's belief that this is best done in a vigorous market economy'. The strategy also identifies problems and opportunities in each sector. It confirmed that the government would continue its policy of green economic instruments.

Organization and Management

The process was run from the office of the Head of Environmental Protection, Central Division, DoE. The 'core team' comprised the head, an editor and two or three staff. A small project team (chaired by the head) was established and met monthly. It comprised departmental staff, professional scientists, economists and statisticians.

A network of government staff throughout all relevant departments worked on various aspects of the strategy and on sections of text, through voluminous correspondence and bilateral meetings (to discuss issues and resolve differences of opinion or perspective). Where issues could not be resolved on a bilateral basis, they were dealt with at progressively more senior levels, and sometimes — where really contentious — by the Cabinet Subcommittee on the Environment.

In November 1992, the government issued an 'open' letter and invited views on the nature of sustainable development and on a possible national strategy. A draft framework for the strategy was then prepared by the DoE and discussed with a number of NGOs and with representatives of EC member states and the European Commission.

In the winter of 1992/3, the DoE held an informal meeting with environmental NGOs on an approach to a strategy, and held limited discussions with the Advisory Committee on Business and the Environment, the Central Local Government Forum, and the Voluntary Sector Environment Forum.

A three-day seminar on sustainable development was held at Green College, Oxford, in March 1993, with about 100 representatives from environmental groups, business organizations, local authorities, academic and research institutions, and government departments. This seminar considered general principles of sustainable development, the implications of UNCED, the scope and content of the national strategy (the DoE presented a draft outline), and policy areas central to sustainability. A summary of the proceedings was prepared (GCCEPU, 1993).

Following the seminar, the DoE elected to change its approach: rather than publishing a bare outline of the strategy as then envisaged, it was decided to adopt a more ambitious approach and to 'fill out' the text. This led to three months of intensive discussions and debate within government and through the GEN 33 committee.

A government consultation paper followed (UKDoE, 1993), providing a more detailed outline of the strategy, and setting out the main topics likely to be covered. It was circulated to over 6000 organizations and individuals. More than 500 responses were received from local authorities, companies (including multinationals and small businesses), voluntary groups, universities, professional institutions and individuals. The strategy contains a summary of these responses.

Coinciding with release of the consultation paper, the *Daily Telegraph* newspaper ran a series of articles with a questionnaire on attitudes and priorities for sustainable development, which generated over 8000 responses.

A 'round-table' process was also initiated with nearly 40 meetings in London involving government departments, NGOs and other interested parties between July and October 1993. This encompassed more than 100 different organizations. Some of these meetings reviewed draft text; others considered issues papers.

Terms of Reference

There were no set terms of reference, apart from a broad mandate given by the Prime Minister who announced the initiative at UNCED. But terms of reference were set out for the various follow-up initiatives (see section below on follow-up).

Reactions

Environmental groups felt that the strategy did not move policy forward as much as they had hoped, or set any new targets for reducing pollution or achieving environmental improvements (other countries, such as the Netherlands, have set quantitative targets on a sector-by-sector basis).

Key 'Assistance' Factors and Problems

The Environment Strategy (HMSO, 1990) provided a foundation. Much had already been decided (i e many principles) and done in developing and following-up on this strategy. The DoE considers that the consultations it undertook were important — 'many NGOs tend to underestimate how much their representations actually helped to shape things'. The establishment of the 'Whitehall network' (i e among government departmental staff) provided a mechanism for real argument and real support.

A 'deadline' had been set to produce the strategy by the end of 1993. While this 'forced the pace', it also placed a limit on how far particular discussions could proceed before a decision was required (i e it cut short some of the debate). The 1993 deadline had been set, in part, because the government wanted 'to set an example to other countries to encourage them to prepare strategies'. The Prime Minister had stressed the need for such strategies at the 1992 meetings of the G7 and of EC heads of government.

Consensus

There was consensus across government on the strategy, and it was signed and presented to Parliament by 16 ministers. Society as a whole is mainly unaware of the strategy. There is no consensus among interest groups. Predictably, environmental groups felt the strategy should have gone much further, while industry was concerned that it should not go too far. However, the publication of the strategy was not seen as an 'end point' and various initiatives for follow-up have been established (see section below on follow-up).

Links to National Planning and Decision-Making

Under the Planning and Compensation Act (1991), 'all local planning authorities are now required to prepare and keep up-to-date a development plan containing policies and proposals relating to the development of the whole of their area, and these must take account of environmental considerations. Development control decisions must normally accord with such plans' (HMSO, 1994a). The government is currently revising its planning policy guidance notes 'in the spirit of UNCED, Agenda 21 and the UK Sustainable Development Strategy'.

A new Environment Agency, bringing all pollution regulators together, came into being in early 1996. It is required, by the 1995 law that established it, to pursue environmental protection through a sustainable development approach. The Secretary of State for the Environment will provide guidance to the new agency on sustainable development.

Regional Strategies

Environmental policy in the UK is now inextricably bound up with European Union policy. Much of the environmental protection legislation is now developed in common with other EU member states, for example, water and air quality, waste management, wildlife and habitats protection, dangerous substances and environmental impact assessment. Parallel to UNCED, the European Commission developed its Fifth Environmental Action Programme, *Towards Sustainability* (CEC, 1992a), which was formally adopted by the Council of Ministers on 1 February 1993 (see Chapter 21). During the development of the UK Sustainable Development Strategy, 'an eye was kept' on the Fifth Environmental Action Programme, but the UK strategy built more on the UK's Environmental Strategy (HMSO, 1990).

Provincial and Local Strategies

In response to Agenda 21 and the European Commission's Fifth Environmental Action Programme, local government in the UK is actively developing its own policies and programmes (parallel to the UK national strategy) through the Local Agenda 21 initiative, which aims to:

- interpret Agenda 21 for local authorities;
- define what sustainable development means at the local level;
- encourage and enable local authorities to prepare strategies for sustainable development;
- develop local environmental initiatives;
- develop information exchange on best practice;
- assist with education and training initiatives; and
- assist with the creation of local partnerships, particularly with the business and voluntary sectors.

Under the initiative, all the local authority associations have come together to set up a Local Agenda 21 steering group, which represents local government, industry, trade unions, the voluntary sector, women's organizations, and higher education. The Local Government Management Board (LGMB) has been nominated to coordinate the Local Agenda 21 initiative, under the direction of the steering group. The LGMB coordinates the following activities:

- publishing guidance, for example step-by-step guides and videos;
- organizing conferences and round-table discussions on aspects of sustainable development at a local level;
- representing the interests of local government in national and international policy; and
- appointing a Local Agenda 21 projects officer to work with local authorities.

A recent survey showed that over 70 per cent of local authorities have now committed themselves to the process, but fewer have been prepared to commit significant financial or staff resources (UNA–UK, 1995). Generally, experience shows that the most progressive authorities are those with a well-established environmental policy and track record in community development work. Another contributing factor to a successful Local Agenda 21 is support and involvement of the staff and elected members. Box 19.1 provides examples of a range of Local Agenda 21 initiatives. Many local authorities have also established links in developing countries (see Box 19.2).

More recently, major non-governmental organizations have become involved in Local Agenda 21s. For example, WWF has developed a model of Local Agenda 21 action based on community education and is piloting this in Reading, and Friends of the Earth has produced a 'Planning for Planet' guide aimed at local authorities, which gives directions on how to use the planning system to achieve sustainable development. However, much of the initial interest has come from the more environmentally-focused groups rather than from other community-related groups.

There is already some evidence of increasing involvement by local community organizations in England and Scotland. The Association of Community Councils in Scotland is running a series of seminars to publicize Agenda 21. It appears that while many such community councils are not aware of the existence of Agenda 21, they are quick to realize its relevance to local action for the environment.

Also, a growing number of businesses are becoming actively involved in Local Agenda 21s. For example, seminars on energy efficiency and waste minimization have been held for small businesses in Luton. However, progress is slow for several reasons:

Box 19.1 EXAMPLES OF LOCAL AGENDA 21 INITIATIVES IN THE UK

Croydon *was one of the first boroughs to employ a full-time Local Agenda 21 officer in July 1993. The council set up five project groups, each dealing with different issues of concern and initiating related projects. Public participation has been limited but the borough now intends to focus on widening community involvement using a variety of innovative ways. It is using a Local Agenda 21 video, prepared by local technicians, to stimulate discussion with community groups and to help involve those groups in the process. 'Croydon Environmental Challenge' has also been set up in partnership with business. For example, British Gas has asked local schools to study the environmental impact of its use of transport and to suggest improvements. Croydon admits that it overlooked the need to set up a mechanism that would facilitate communication between groups and, in retrospect, this should have been a priority.*

Gloucestershire *decided it was essential to involve the community sector in the Local Agenda 21 process and asked a local charity, 'Rendezvous', to run the process — this arrangement is unique in the UK. Through extensive networking, the coordinator of Rendezvous managed to enlist considerable support for the process. The United Nations Association (UNA) was engaged as an adviser and a separate training budget was set up. The process was retitled 'Vision 21' and eight working groups were established on energy, waste, transport, social and health issues, natural resources, the economy, the built environment, and education and community involvement. These groups are drawn together by a coordinating group with two further linking groups: an LA21 linking group and an education group that has become the Education and Community Involvement Network (ECIN). The second phase of Vision 21 will involve a community-wide 'visioning' exercise where the initial analysis of issues and problems will lead to the development of strategies which will, in turn, form part of the draft Local Agenda 21 document. It has been realized that, to take this document forwards, will need considerable resources, a business plan and a fundraising strategy.*

Mendip District Council *started an environmental programme in 1991 inspired by the Friends of the Earth local government charter. In 1992, the council undertook an environmental audit of its activities. This brought about a change in council policies and the production of a 'corporate strategy'. Subsequent measures have included the sending of staff on 'green awareness' training programmes, the development of a green purchasing policy, and a requirement for the environmental appraisal of all applications. An Environment Forum was set up in 1993 which, together with the council, has initiated several projects. These include: the 'Age to Age' initiative which brings young and old people together to talk about the past; 'A Map of Mendip' which combines information supplied by the community with the council's GIS. The council and forum are now planning workshops to help promote Agenda 21, focusing around themes such as women and environment, business and environment, and farming and environment. Health professionals can also feed into the LA21 process through the Health Alliance.*

Nottinghamshire's *LA21 process has benefited from a high degree of cooperation* *between its nine local authorities. Although progress has been varied, slower authorities have been quick to catch up and each authority has its own approach to developing and implementing LA21. Rushcliffe District Council's LA21 has been particularly successful in involving schools, Bassetlaw District Council has maximized participation by rotating venues and chairpersons and by having the agendas set by different sectors. It has also started on the implementation of the Eco-Management and Audit Scheme (EMAS) and has initiated some small-scale demonstration projects, but adoption has so far been limited. Other districts are less advanced. The county has concentrated on 'Greening the Council' and 'Greening the Community' strategies which address some LA21 objectives. A 'Greening Your School' pack is in production. An 'Officer Joint Working Party on Green Issues' enables officers dealing with LA21 issues in all the councils to meet regularly. This group has also run seminars on sustainable development for economic development officers. There are several joint projects such as the establishment of composting trials to assess domestic waste reduction. Two youth environment fora have been organized and the county has also trained youth workers on environmental issues and produced a green audit system for youth clubs. 'It's our Environment Too' environmental awareness initiatives have been organized in black and ethnic communities. The East Midlands Business and Environment Club (which also covers four other counties) has been set up. About half of its 90 representatives come from small to medium sized enterprises which previously have been reluctant to participate in such activities.*

Plymouth *began its LA21 process by setting up an Environment Forum, which has since developed a separate identity and is supported by all major agencies in the city. The forum meets bimonthly and an atmosphere of trust and open discussion has developed which has facilitated the development of crosscutting work on issues such as asthma and air pollution. Local neighbourhoods are writing their own community action plans, a young persons' conference has been organized, and an Environmental Resource Centre is being set up. The challenge now remains to maintain the enthusiasm and the focus has moved away from practical projects bringing tangible benefits to the equally important 'intangibles' such as improving communications between different sectors. The LA21 process in Plymouth, as in many parts of the UK, sees education and young people as priority areas.*

Reading *has received considerable assistance from WWF in the LA21 process, building on its work on community education. There are three main areas of work: a State Of Environment Report, A Business Agenda 21 initiative and neighbourhood Agenda 21. The latter initiative has led to the formation of an organization named GLOBE (Go Local for a Better Environment), which has helped set up a series of community consultation meetings in two pilot neighbourhoods. Emphasis was placed on the identification of local solutions rather than the provision of a 'wish list' for the council. GLOBE is now supported by WWF and it is hoped that local businesses will assist with further funding. Neighbourhoods have gradually 'warmed' to the initiative as they realize that implementation of solutions depends on cooperation between other communities and neighbourhoods. A twinning project between Reading and Tanzania is also being developed with assistance from WWF. Local strategies developed under GLOBE will feed into the council's own action plans and LA21.*

Vale Royal *is a borough centred on two towns in the county of Cheshire, Northwhich and Winsford. The council developed an environmental charter in 1992 and has since established a strong working relationship with the Vale Royal Environment Network (VREN). The Director of Environmental Services and Housing, who is also the elected leader of the council and director of the local Council for Voluntary Service has, through his own interest, become involved in the process. Great effort was made to ensure maximum participation in LA21 seminars by running seminars twice, and the LA21 support group decided to take on a full-time LA21 officer. Since her appointment, several events have been organized, including a Cheshire-wide conference on Agenda 21. Round tables on eight themes have been set up, each with a facilitator and designated staff member. The Chamber of Commerce is trying to encourage a local business person to take part in each one. Several planned joint initiatives have emerged out of the round-table meetings and community activities have also taken place. For example, school sixth formers around Cheshire produced their own Agenda 21, and a branch of the National Council for Women brought together local women to discuss their perspectives. Vale Royal admits that coordination of so many events would not be possible without a paid coordinator, but the momentum has continued to grow and, with continued interest of senior management and staff, the 'Vision for Vale Royal' may achieve its ambitious goals.*

the recession makes it harder to get financing for longer-term investments; there is a lack of clear leadership in the business community; and public scrutiny continues.

During the last decade there has been a large increase in environmental activity within local authorities, and Local Agenda 21 is not entirely new. But it would be erroneous to subsume Local Agenda 21 under a purely environmental banner. Considerable progress has been made with organizing consultation and participation, with great emphasis placed on giving individuals and organizations an opportunity to contribute actively to the process. Interestingly, several local authorities have begun to develop alternative titles for Local Agenda 21 (for example, 'A Vision for Vale Royal', 'A Blueprint for Leicester'), but others feel the original name should continue to be used.

Convention-Related Strategies

During the preparation of the UK Sustainable Development Strategy (HMSO, 1994a), the UK Programme for Climate Change (HMSO, 1994b) and the UK Biodiversity Action Plan (HMSO, 1994c) were developed. A UK Programme for Sustainable Forestry was also produced (HMSO, 1994d). These exercises were conducted independently of the sustainable development strategy, but with mutual feedback between the various processes. All were published together on 25 January 1994 and the UK Sustainable Development Strategy contains the main findings from the other three.

Ecological Footprints and Transboundary Issues

The strategy raises the concept of 'ecological footprints' as a 'difficult issue to think about', but without resolving the government's view. It has taken a 'cautious position'. Cross-boundary issues are discussed with regard to Europe and mainly in terms of

Box 19.2 NORTH–SOUTH LINKAGES AND LOCAL AGENDA 21 IN THE UK

In 1993, under the Local Government Act, local authorities were given new legislative powers to participate in overseas assistance programmes. This has enabled local government to collaborate with southern partners in the promotion of policies and programmes that promote sustainable development practice. A large number of authorities in the UK have already established international links and it has been recognized that such links could make a significant contribution to the Local Agenda 21 process. These links could be developed further to promote strategies for sustainable development and safeguarding the environment in local communities. They could also be a basis for partnership in the search for sustainable solutions at the local level. North–South links have provided a means for direct contact with other cultures, understanding of North–South priorities, constraints and concerns.

Activities have included:

- *support to grass-roots community-based initiatives in linked overseas communities;*
- *participation in joint discussions/round tables to identify sustainable development priorities, and agree on joint activities;*
- *facilitation of UK-based work experience and training for overseas personnel;*
- *organization of professional development programmes, which have included overseas visits by UK local government personnel/technical specialists, exchange programmes, and secondment of staff to overseas organizations concerned with the functioning of local government;*
- *promotion of education, awareness raising and development education activities;*
- *support and facilitation of joint ventures and sustainable local development project initiatives overseas; and*
- *encouragement and promotion of community fund-raising initiatives in support of overseas projects.*

Examples of North–South links include:

Oxford City Council, *which has had an established link with Léon City Council since 1985. Presently, the major project being supported is the Léon City Council 'Rio Chiquito Project', which aims at clearing up the heavy pollution in the river flowing through a heavily populated area of Léon city. The project is attempting to raise US$ 6 million to improve local water supply, tree planting programmes, improved housing, roads and waste disposal.*

Calderdale Metropolitan Borough Council, *which recently developed a link with Musoma in Tanzania. The community and civic relationship is being built parallel to an existing diocesan link. Activities have included: a visit to Tanzania by a Calderdale GP to provide advice on health and environmental health issues; attendance of an officer from the Musoma City Council at a management course at Huddersfield University; and support to a local blind school in Tanzania. School to school links are being encouraged. Calderdale Council has a part-time twinning officer who is also coordinating the link with Calderdale and a new link in the Czech Republic.*

The UK Local Government International Bureau (LGIB), *which promotes international cooperation in local government and provides advice on North–South linking and initiatives in Europe. LGIB, in collaboration with the Local Government Management Board, has been involved in various capacity-building measures for local government. Activities have included a workshop held in 1992 at the Government Training Institute in Kenya, under the auspices of UK ODA, to strengthen the institution's ability to take on a role as a regional local government training centre; a senior policy seminar on decentralization and strengthening local government in Africa in The Gambia in 1992; and an event entitled 'Strategies for Decentralized Government' was held in Harare and Lagos and sponsored by the Commonwealth Secretariat.*

The International Council of Local Environmental Initiatives (ICLEI), *which coordinates a worldwide Local Agenda 21 programme and encourages North–South linkages. ICLEI is a democratic organization governed by an executive committee comprising 21 individuals elected from its membership and representing all regions of the world. ICLEI Local Agenda 21 initiatives incorporate the Local Agenda 21 Model Communities Programme, which aims to develop tools and models of sustainable development planning, and the Local Agenda 21 Communities Network, which is a larger network of local governments and their partners undertaking sustainable development planning processes. ICLEI is affiliated to the International Union of Local Authorities (a worldwide association of local government bodies with a commitment to the promotion of sustainable development and programmes of intermunicipal cooperation). It has regional offices in Canada, Africa and South America.*

UK-based Organizations

Local Government International Bureau, Development Cooperation Section, 35 Great Smith St, London SW1P 3BJ (Tel: 0171 222 1636. Fax: 0171 233 2179) *promotes international cooperation in local government and provides advice on North–South linking and initiatives in Europe. Provides information on related organizations.*

Local Government Management Board, Environment Adviser or Local Agenda 21 Project Officer, Arndale House, The Arndale Centre, Luton, LU1 2TS (Tel: 01582 451166. Fax: 01582 412525) *helps coordinate the UK Local Agenda 21 initiative and any associated environment initiatives.*

International Union of Local Authorities, The Secretary-General, PO Box 90646, 2509, LP The Hague, Netherlands (Tel: +31 70 32 44 032. Fax: +31 70 32 46 916). *A worldwide association of local government bodies with a commitment to the promotion of sustainable development and programmes of intermunicipal cooperation.*

Commonwealth Local Government Forum, 35 Great Smith Street, London SW1P 3BJ (Tel: +44 171 799 1730. Fax: +44 171 799 1731). *The forum was launched on 23 March 1994 and an important aspect of CLGF's work is 'to reflect current trends in democratic and administrative reform with a view to disseminating the best prevailing practice, while recognizing the varying needs of diverse local communities'.*

International Council for Local Environmental Initiatives, The Secretary-General, ICLEI Secretariat, 8th Floor, East Tower, City Hall, 100 Queen St West, Toronto, ON M5H 2n2, Canada (Tel: +1 416 392 1462. Fax: +1 416 392 1478). *As mentioned above, with its 21-person executive committee elected from among its membership and all regions of the world being represented, ICLEI is a highly democratic organization set up to coordinate a worldwide Local Agenda 21 programme and encourage North–South linkages. For its Local Agenda 21 initiatives, it makes use of both the Local Agenda 21 Model Communities Programme, designed to develop tools and models for sustainable development planning, and the Local Agenda 21 Communities Network, which is a larger network of local governments and their partners undertaking sustainable development planning processes.*

Towns and Development, The Coordinator, PO Box 85615, 2508 CH The Hague, Netherlands (Tel: +31 70 362 3894. Fax: +31 70 364 2869). *This is an international consortium of NGOs and local government organizations promoting local initiatives for sustainable development. T&D aims to encourage local authorities, NGOs and community-based groups to work together for North/South cooperation.*

'shared' media, for example, the air and the sea. There were no negotiations on cross-boundary issues (other than in the course of international negotiations on specific issues, which the strategy then reflected).

Parliamentary Process

The strategy was submitted to Parliament by 16 ministers (see section above on initiation and time perspective) and there were references to it in the House of Commons but no set piece debate. In 1994 and 1995, the House of Commons Select Committee on the Environment questioned the Secretary of State for the Environment on progress with the strategy when considering the annual progress reports. The 1994 report (HMSO, 1994a) included a résumé of the strategy while the 1995 report (HMSO, 1995) reported on progress in implementation and follow-up.

A temporary House of Lords (upper chamber of Parliament) Select Committee on Sustainable Development was established in the Spring of 1994. It took evidence, interviewing expert witnesses, and produced a report in July 1995. The House of Lords debated the report and the government's response in October 1995.

Follow-Up

The UK Sustainable Development Strategy has a 20-year perspective and there are no formal plans or commitments to revise or update it — this will depend on the government of the day. However, the DoE would like to be able to review the strategy after a few years (five or six). But, in order to monitor commitments openly, an annual progress report is continuing to be produced (and is now combining reporting on the 1990 Environmental Strategy and the 1994 Sustainable Development Strategy).

While the Strategy did not contain new economic instruments, 'it prepared the way for separate decisions that did' (Stevens, 1995). The 1995 budget introduced a new tax on the disposal of waste in landfill sites — a measure to encourage recycling and waste minimization — and combined this with reductions in taxation on employment. A new 'packaging initiative' has been introduced which requires industry to make proposals for a self-run levy scheme.

The government remains committed to the principle of annually increasing duty on transport fuel by at least an average of 5 per cent above the rate of inflation per year, and to introducing VAT on energy consumed in the home. However, the second of these measures met with opposition in Parliament. The rate of VAT was therefore left at 8 per cent (after the first stage was implemented) — rather than being raised to 17.5 per cent (the second stage) as planned. In lieu of this, other measures were taken to keep the CO_2 programme on course.

The strategy also identified the need for new indicators of sustainable development. A task force was established, with members from all the main ministries, and proposals were expected by the end of 1995. In parallel, local authorities are also developing their own sustainability indicators.

Apart from the climate, biodiversity and forestry programmes/plans, the government has set out new national targets for reducing some pollutants and making proposals on others. In addition, a draft strategy for dealing with waste has been published, containing quantifiable targets, for example, reduction of waste going to landfill by 10 per cent over the next ten years and a further 10 per cent in the following ten years. Consultations on this draft strategy have been held.

Following publication of the sustainable development strategy, three new bodies have been established:

Government's Panel on Sustainable Development — consists of five eminent people in Britain, appointed by the Prime Minister, with responsibility to advise the government and monitor its performance. The terms of reference for the panel set out its remit as:

- to keep in view general sustainability issues at home and abroad;
- to identify major problems or opportunities likely to arise;
- to monitor progress; and
- to consider questions of priority.

Panel members have direct access to all ministers and civil servants. During their first year they called for papers on a number of subjects and gave advice in private. After one year, an open report to the Prime Minister was published.

UK Round Table on Sustainable Development — a body of about 35 people from all sectors and groups, chaired jointly by the Secretary of State for Environment and an ex-Vice-Chancellor of Oxford University. In an answer to a parliamentary question on 19 January 1995, the Secretary of State for the Environment outlined the Round Tables' objectives as:

- to help identify the agenda and priorities for sustainable development;
- to develop new areas of consensus on difficult issues of sustainable development and, where this is not possible, to clarify and reduce difference;

- to provide advice and recommendations on actions to achieve sustainable development;
- to help evaluate progress towards objectives; and
- to inform and involve others, building wider support for emerging consensus.

This body took a year of difficult consultation and negotiations to get going, but has now had four meetings. Green ministers have been interviewed and agreement has been reached to study transport, economic instruments, environmental appraisal and energy.

Going for Green — a new campaign to carry the message of sustainable development to private individuals, run by a committee of prominent people from different walks of life. The committee will work with existing groups, but give new publicity to issues and establish new initiatives.

Chapter 20 UNITED STATES of AMERICA

The President's Council on Sustainable Development (PCSD) has recently presented to the President a national sustainable development action strategy, *Sustainable America: A New Consensus for Prosperity, Opportunity, and a Healthy Environment for the Future* (PCSD, 1996). This report recommends innovative changes in the nation's regulatory, policy and community decision-making processes — for both institutions and individuals — in order to achieve the complementary goals of economic prosperity, environmental protection and greater social equity. The work of the PCSD has been the only national-level strategy process, and PCSD staff interviewed talked of its work as 'multi-stakeholder'. The Environmental Protection Agency (EPA) is working on an 'Environmental Goals Project', initiated in 1992 (see section below on focus), which also involves a public participation process. Various state and city strategies for sustainable development have been developed (for example, Minnesota, Detroit, Seattle, Chicago, Chattanooga). Many of these have been catalogued by the Centre for Policy Alternatives (Nguyen and Roberts, 1994).

The PCSD work involved no formal linkage with efforts to develop state or city strategies, although individuals involved in the latter have participated in task force meetings and attended and spoken at council meetings, and the PCSD tried to 'capture' state and local-level experience; and there was a dialogue with local, county and state governments to coordinate and catalyse synergy.

The USA has signed the Biodiversity Convention but has not yet ratified the treaty, and has no plans for a national biodiversity strategy at present. Individuals from EPA and the Department of Energy who were involved in preparing the US climate action plan (mainly a packaging of voluntary initiatives) were involved in the work of the PCSD.

Numerous US government agencies are working on sustainable development initiatives, for example, the integration of economic, environmental and social policies; mitigating the effects of depleted fish stocks caused by overfishing and man-induced factors; and exploring opportunities for demonstration projects to exemplify sustainability in action.

President's Council on Sustainable Development

Initiation and Time Perspective

As part of the work of the President's Commission on Environmental Quality (PCEQ) (established by the Bush administration), Frank Popoff of the Dow Chemical Company headed a task force to develop a model for international discussion on sustainable development. It was thought that such discussion was needed in the USA before an international forum was convened. This idea was pursued by the PCEQ members and with the incoming Clinton administration.

The PCSD was subsequently established on 29 June 1993 by President Clinton through Executive Order 12852 (Box 20.1). Its members were appointed by the President from the public and private sectors representing industrial, environmental, governmental and not-for-profit organizations with experience related to sustainable development. The tasks of the PCSD were:

- to develop and recommend to the President a national sustainable development action strategy that will foster economic vitality;
- to develop an annual Presidential Honors Program recognizing outstanding achievements in sustainable development; and
- to raise public awareness of sustainable development issues and participation in opportunities for sustainable development.

The PCSD had a two-year term. Council meetings were quarterly, with alternate ones held outside Washington DC. The PCSD prepared a draft strategy report for public comment in the Autumn of 1995. It included goals, indicators of progress and policy recommendations. The final report was published and presented to the President in February 1996.

Prime Motivation

Several important factors influenced the PCSD initiative:

- the controversy in the northwest Pacific over jobs versus the environment;
- President Clinton's disposition to addressing the issue of integrating the environment and the economy;
- the influence of Vice-President Gore, who has a track record of interest in environmental issues;
- the need for a response to commitments entered into by the Clinton administration in response to UNCED;
- interest in a multi-stakeholder, consensus model of development; and
- the Vice-President's ideas and efforts to save money, cut costs and streamline processes within the federal government under the 'reinventing government' initiative.

Box 20.1 EXECUTIVE ORDER NO 12852

By the authority vested in me as President by the Constitution and the laws of the United States of America, including section 301 of title 3, United States Code, it is hereby ordered as follows:

Section 1. Establishment

There is established the 'President's Council on Sustainable Development ("Council")'. The Council shall consist of not more than 25 members to be appointed by the President from the public and private sectors and who represent industrial, environmental, governmental, and not-for-profit organizations with experience relating to matters of sustainable development. The President shall designate from among the Council members such official or officials as he shall deem appropriate. The Council shall coordinate with and report to such officials of the Executive Branch as the President or the Director of the White House Office on Environmental Policy, shall from time to time determine.

Section 2. Functions

(a) The Council shall advise the President on matters involving sustainable development. 'Sustainable development' is broadly defined as economic growth that will benefit present and future generations without detrimentally affecting the resources or biological systems of the planet.

(b) The Council shall develop and recommend to the President a national sustainable development action strategy that will foster economic vitality.

(c) The chairperson or chairpersons may, from time to time, invite experts to submit information to the Council and may form subcommittees of the Council to review and report to the Council on the development of national and local sustainable development plans.

Section 3. Administration

(a) The heads of executive agencies shall, to the extent permitted by law, provide to the Council such information with respect to sustainable development as the Council requires to carry out its functions.

(b) Members of the Council shall serve without compensation, but shall be allowed travel expenses, including per diem in lieu of subsistence, as authorized by law for persons serving in the government service (5 USC 5701–5707).

(c) The White House Office on Environmental Policy shall obtain funding for the Council from the Department of the Interior or such other sources (including the Federal agencies) as may lawfully contribute to such activities. The funding received shall provide for the administration and financial support of the Council.

(d) The Office of Administration in the Executive Office of the President shall, on a reimbursable basis, provide such administrative services for the Council as may be required.

Section 4. General

(a) I have determined that the Council shall be established in compliance with the Federal Advisory Committee Act, as amended (5 USC App). Notwithstanding

any other Executive Order, the functions of the President under the Federal Advisory Committee Act, as amended, except that of reporting to the Congress, which are applicable to the Council, shall be performed by the Office of Administration in the Executive Office of the President in accord with the guidelines and procedures established by the Administrator of General Services.

(b) The Council shall exist for a period of 2 years from the date of this order, unless the Council's charter is subsequently extended.

(c) Executive Order No 12737, which established the President's Commission on Environmental Quality, is revoked.

William Clinton

The White House
June 29, 1993

AMENDMENT TO EXECUTIVE ORDER No 12852

By the authority vested in me as President by the Constitution and the laws of the United States of America, including section 301 of title 3, United States Code, and in order to amend Executive Order No 12852, it is hereby ordered that Executive Order No 12852 is amended by deleting the text of Section 3(d) of the Order and inserting in lieu thereof the following text: 'The Department of the Interior shall, on a reimbursable basis, provide such administrative services for the Council as may be required' and by deleting the words 'Office of Administration in the Executive Office of the President' in Section 4 of that order and inserting the 'Department of the Interior' in lieu thereof.

William Clinton

The White House
July 19, 1993.

Focus

At the first meeting of the PCSD in July, 1993, it adopted the definition of sustainable development offered by the Brundtland Commission (WCED, 1987). Through the course of its work, the PCSD developed and released several working documents. They included a *Challenge Statement* and a draft *Vision Statement and Principles of Sustainable Development*. Public comment was solicited on the latter document and nearly 800 individuals and organizations responded (Bear, 1994). The statement was revised and a further draft approved by the PCSD in October 1994. Parts of a draft strategy were released in early August 1995 for further public comment. The final report, on which there was unanimous agreement among PCSD members, includes:

- a 'vision statement' articulating the PCSD's broad concept of the benefits of sustainability for the nation;
- a 'we believe statement' setting out fundamental beliefs on sustainability;
- ten 'national goals' expressing, in concrete terms, the elements of sustainability;

and suggested indicators to measure progress towards achieving them (see Box 20.2);

- 154 specific actions to carry out 38 major policy recommendations in several areas — building a new framework for a new century; information and education; strengthening communities; natural resources stewardship; population and sustainability; and international leadership.

In a 'highlights' leaflet issued along with the published final report, the PCSD states that:

> *At the heart of the Council's recommendations is the conviction that economic, environmental and social equity issues are inextricably linked and must be considered together. To achieve sustainability, institutions and individuals must adopt this new way of thinking. Another core belief is that to achieve sustainability, some things — jobs, productivity, wages, capital and savings, profits, information, knowledge, and education — must grow, and others — pollution, waste and poverty — must not.*
>
> *Government decision-making must provide direct and meaningful interaction among those affected. Accommodation by all levels of government and vast technological changes in information delivery must occur to enable citizens and institutions to participate more fully. New methods of governance that are more collaborative, such as private partnerships, must be instituted. For example, in the area of environmental management, the Council's recommendation to create a new regulatory system would require federal, state and local governments to work with business and citizen groups to develop strategies applicable to local areas. Moreover, the strategies would have to integrate economic, social and environmental issues. Both individuals and institutions must take on greater responsibility, and embrace an ethic of environmental stewardship.*

While the PCSD was, to some extent, wrestling with the international component, the focus of its work was primarily domestic and it was unable to invest the time and resources to deal effectively with the concept of 'ecological footprints'. Foreign policy and foreign aid issues were mainly set aside. Trans-boundary issues were not dealt with explicitly in documentation, but may have been discussed by task forces.

Organization and Management

The PCSD comprised 25 members with two co-chairmen — David T Buzzelli, Vice-President of the Dow Chemical Company; and Jonathan Lash, President of the World Resources Institute. There were five Cabinet secretaries (Energy, Commerce, Industry, Agriculture and the Environmental Protection Agency) and four government Deputy/Under-Secretaries served as ex-officio members. Council members served on eight main task forces:

- eco-efficiency;
- energy and transportation;
- natural resources management and protection;

Box 20.2 US NATIONAL GOALS TOWARDS SUSTAINABLE DEVELOPMENT AND EXAMPLE INDICATORS OF PROGRESS

Goal 1 Health and the Environment *Ensure that every person enjoys the benefits of clean air, clean water and a healthy environment at home, work and play.*

- ***Clean air*** *A decrease in the number of people living in areas that fail to meet air quality standards.*
- ***Diseases and mortality*** *A decrease in diseases and deaths from environmental exposures, including occupation related illnesses.*

Goal 2 Economic Prosperity *Sustain a healthy US economy that grows sufficiently to create meaningful jobs, reduce poverty and provide the opportunity for a high quality of life for all in an increasingly competitive world.*

- ***Employment*** *Increase the number, wage level and quality of jobs (as measured, for example, by the percentage of jobs at or below minimum wage).*
- ***Poverty*** *Decrease the number of people living below the poverty line.*

Goal 3 Equity *Ensure that all Americans are afforded justice and have the opportunity to achieve economic, environmental and social well-being.*

- ***Environmental equity*** *Development of any disproportionate environmental burdens (such as exposure to air, water and toxic pollution) borne by different economic and social groups.*
- ***Social equity*** *Development of measures of access to critical services (such as education, health care and community services), and opportunities to participate in decision-making by different economic and social groups, such as the percentage of these populations attending college.*

Goal 4 Conservation of Nature *Use, conserve, protect and restore natural resources — land, air, water and biodiversity — in ways that help ensure long-term social, economic and environmental benefits for ourselves and future generations.*

- ***Habitat loss*** *Development of measures of threats to habitat loss and the extent of habitat conversion, such as the rate of wetlands loss.*
- ***Ecosystems*** *Increased percentage of forests managed to reach full maturity and diversity.*

Goal 5 Stewardship *Create a widely held ethic of stewardship that strongly encourages individuals, institutions and corporations to take full responsibility for the economic, environmental and social consequences of their actions.*

- ***Waste reduction*** *Increased source reduction, reuse, recovery and recycling.*
- ***Energy efficiency*** *Reduced energy intensity (energy per unit output).*

Goal 6 Sustainable Communities *Encourage people to work together to create healthy communities where natural and historic resources are preserved, jobs are available, sprawl is contained, neighbourhoods are secure, education is lifelong,* *transportation and health care are accessible, and all citizens have opportunities to improve the quality of their lives.*

- ***Safe neighbourhoods*** *Decrease in violent crime rates.*
- ***Shelter*** *Decreased number of homeless people by community.*

Goal 7 Civic Engagement *Create full opportunity for citizens, businesses and communities to participate in and influence the natural resource, environmental and economic decisions that affect them.*

- ***Public participation*** *Increase in the percentage of eligible voters who cast ballots in national, state and local elections.*
- ***Citizen participation*** *Increase in community participation in such civic activities as professional and service organizations, parent-teacher associations, sporting leagues, and volunteer work.*

Goal 8 Population *Move toward stabilization of US population.*

- ***Population growth*** *Reduced rate of population growth in the USA and world.*
- ***Status of women*** *Increased educational opportunity for women; increased income equality for equivalent work.*

Goal 9 International Responsibility *Take a leadership role in the development and implementation of global sustainable development policies, standards of conduct, and trade and foreign policies that further the achievement of sustainability.*

- ***International assistance*** *Increased level of US international assistance for sustainable development, including official development assistance (federal money).*
- ***Environmental technology exports*** *Increased US exports or transfers of cost-effective and environmentally sound technologies to developing countries.*

Goal 10 Education *Ensure that all Americans have equal access to education and lifelong learning opportunities that will prepare them for meaningful work, a high quality of life, and an understanding of the concepts involved in sustainable development.*

- ***Information access*** *Increased number of communities with infrastructure in place that allows easy access to government information, public and private research, and community right-to-know documents.*
- ***National standards*** *Increased number of school systems that have adopted K–12 voluntary standards for learning about sustainable development similar to the standards developed under the National Goals 2000 initiative.*

Source: PCSD, 1996.

- principle's, goals and definitions;
- population and consumption;
- public linkage, dialogue and education;
- sustainable agriculture; and
- sustainable communities.

Each task force expanded its capacity by bringing in other experts and sectoral representatives, who sometimes operated as members of working groups, or of subject or regional teams. These variously conducted scenario planning; carried out research, interviews, field visits and surveys for analysis and discussion; conducted workshops and round-table discussions; undertook case studies; and developed networks — all for the purpose of developing goals and recommendations for consideration by the Council. The task force work was completed in July 1995.

The Council and task forces were serviced by a small secretariat headed by an executive director. The Council met quarterly in public. The 1993 meetings were held in Washington. Subsequent meetings were held alternately in Washington and cities around the country (Seattle, Chicago, Chattanooga, and San Francisco).

Figure 20.1 illustrates the key steps of the PCSD process.

Participation

All Council meetings were open to the public. On average, each was attended by 200–500 people. Some draft documents were released to the public and were available on line. The secretariat produced a quarterly information pack and a newsletter which was distributed to about 5000 people. Some 450 experts were involved in working group activities related to the work of the PCSD and its task forces. On average, about 100 meetings of task forces and their working groups were held each quarter. The draft vision statement and principles on sustainable development were commented on by nearly 1000 individuals and organizations.

Each task force operated differently. They were heavily driven by their participants — the stakeholders set the priorities and developed recommendations for the full Council. But ultimately, the Council made final decisions and developed the final report to the President. Task forces adopted different processes and work places to match the nature of their scope of work. For example, the Sustainable Communities Task Force decided to emphasize a 'grass-roots' approach and involve activists in this movement in all its deliberations. Council meetings or workshops were held in communities that had done pioneering work to realize 'sustainability' in their social, economic and environmental life.

The Natural Resources Task Force held workshops and case studies on watersheds across the country, and members listened to local and regional stakeholders share views on the importance of natural resources and watersheds and share their views for a sustainable future.

There was a reasonably good level of involvement of national-level NGOs, but minimal input from grass-roots NGOs. While the latter appear to have supported the concept of the PCSD, they were concerned that this might result in 'unwinding' some of the hard-won regulatory mechanisms and legislation and eliminating some environmental protection measures. They were also reluctant to become fully engaged because of the high demand in cost and time terms.

Some PCSD task forces were able to support travel costs (but not salaries/fees) of NGOs and grass-roots organizations. By comparison, industry could 'afford' to be involved. A strong effort to maintain a balance was made by the co-chairs and the executive director.

In its final report, the PCSD recommends a suite of actions to 'create a community-

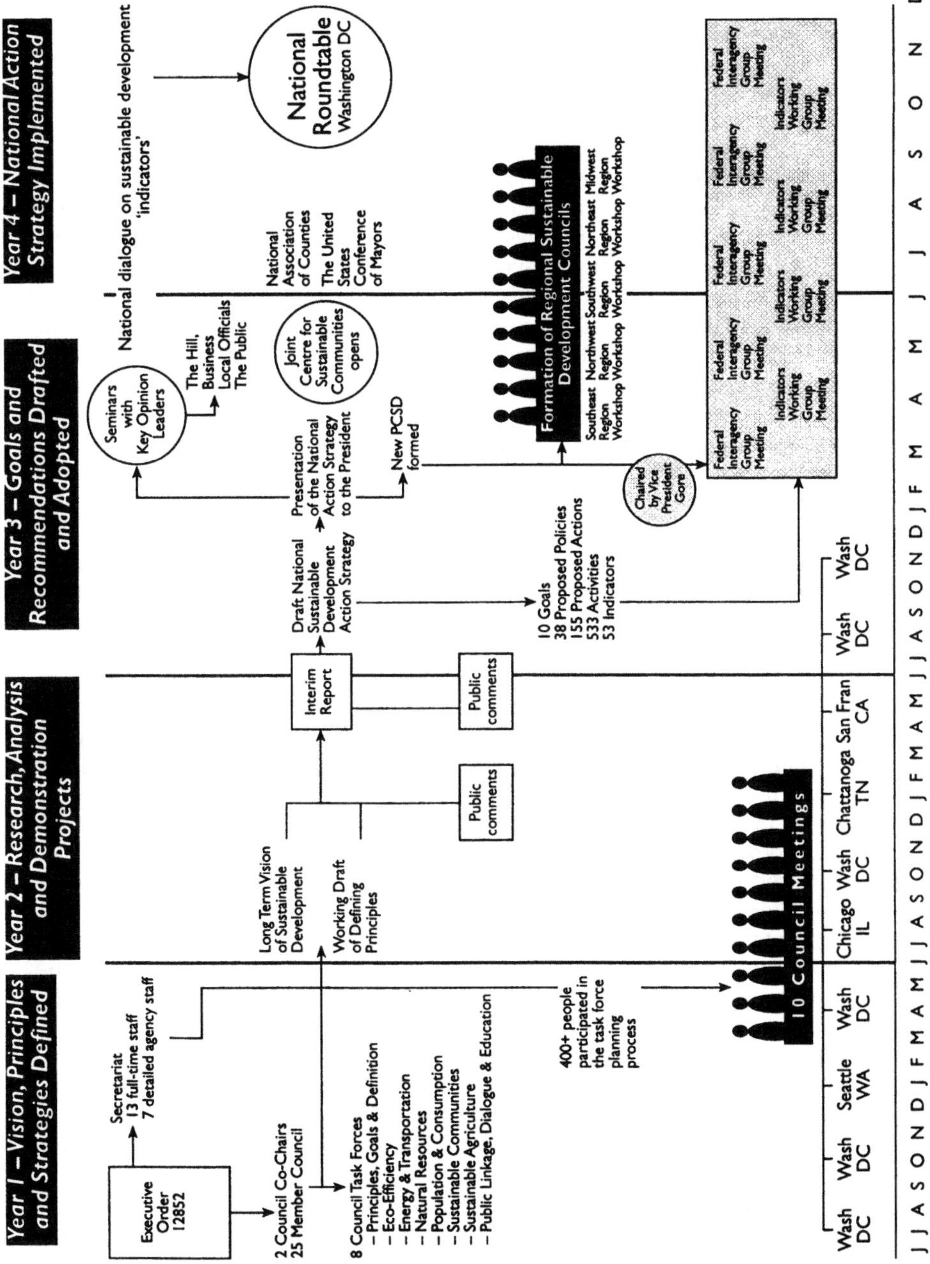

Figure 20.1 President's Council on Sustainable Development: Key Steps to the Submission of the National Sustainable Development Strategy

driven strategic planning process that brings people together to identify key issues, develop a vision, set goals and benchmarks, and determine actions to improve their community' (see Box 20.3).

Key 'Assistance' Factors, Problems and Issues

Former President Bush was judged to have played an unhelpful role at UNCED (refusing to sign the accords). Key public players judged there to be a need for a national sustainable development strategy in the USA. One of President Clinton's first acts was to sign up to Agenda 21. Vice-President Gore also played an influential role.

A number of departments and companies had already launched their own sustainability initiatives. For example, the Department of Commerce was developing a green GDP and issued a sustainable development plan, and the Environmental Protection Agency had summarized its sustainable development activities and initiated a National Environmental Goals Project. Furthermore, members of the Business Council for Sustainable Development in the USA were very supportive.

The work of the PCSD was restrained by the two-year time frame and by limited financial resources. Council members comprised many chief executive officers of government and corporate organizations, and NGOs, who, for their own reasons, all tended to function myopically. The process was also a learning curve for them. A major challenge was for the PCSD to resist involving itself in current political/legislative activities and to train itself to take a longer-term view. The change of Congress in January 1995 was also a factor which that progress.

The Federal Advisory Committee Act required Council meetings to be held in public. While a virtue, this transparency and openness meant that Council members tended to put their own 'party line' at meetings. It slowed the development of confidence, credibility and trust and a preparedness to be candid. 'They cannot be vulnerable in a way that is necessary. Members tended to keep quiet where they had no direct interest, and they were reluctant to speak on issues where they did not have any familiarity or knowledge.' Another problem was that a number of key agencies, while represented, did not give their wholehearted support. The PCSD was a new initiative and it took time (which was in any case limited) to gain respect. There are pros and cons in trying to address the complex issue of sustainable development in the USA. The PCSD could bring key actors to the 'table', but could not eliminate cynicism, parochialism and political rivalry.

The USA has a culture and history of confrontational policies on the environment and of adversarial behaviour in, for example, trade versus the environment. But there has also been an evolution towards willingness to participate in collaborative processes. For example, the good consensus reached on wetland management and hazardous waste control influenced legislation. The PCSD was an attempt to build on this trend and reach consensus on large-scale issues.

The USA has approached environmental protection 'entirely anecdotally'. There are 3000 pages of statutes and 12,000 pages of regulations at the federal level. In each case, there has been a narrow, command and control, reactive response to a specific problem that received public attention. Rarely have these responses and regulations been linked together. This is consistent with the prevailing antagonism towards planning in the USA. There is broad acceptance that it is harder to proceed further through more

Box 20.3 RECOMMENDED ACTIONS TO CREATE A COMMUNITY-DRIVEN PLANNING PROCESS IN THE USA

(1) *All levels of government and the private sector should build a multisector decision-making capacity at the local level. They can do so by providing information and financial and technical assistance to communities that wish to engage in a collaborative, community-wide process to integrate economic prosperity, environmental health, and opportunity in their decisions and actions.*

(2) *All levels of government should ensure substantial opportunity for public participation in all phases of planning and decision-making to allow those affected to have a voice in the outcome. Specific steps include creating and expanding methods for public participation in legislation, ordinances and community advisory boards. Special steps should be taken to ensure that historically under-represented groups are involved.*

(3) *All levels of government, especially local government, should identify barriers to greater citizen involvement in decision-making — such as lack of child care or transportation — and develop strategies to overcome them. Employers should give employees flexibility and incentives to increase the time they and their families can devote to community activities.*

(4) *Community-based coalitions can create educational media campaigns to encourage citizen participation in government, disseminate high-quality information on community issues, and promote public discussions that identify solutions to problems. Coalitions should be as broad as possible, including industry and business, schools, newspapers, television and radio stations, community groups, environmental organizations, labour and local government.*

(5) *Federal and state agencies should help local communities that wish to use profiles of potential environmental risks as a tool to identify and set priorities for solving environmental problems. The agencies should provide information on and facilitate access to communities that have successfully used this tool.*

(6) *Community-based coalitions can work together to draft an economic development strategy to fill basic needs and take advantage of new trends as part of the strategic planning process. Coalitions should include businesses, employees, unions, chambers of commerce, environmental organizations, local government and residents.*

(7) *Community-based coalitions can develop and carry out programmes to increase voter registration and participation, working with national voter registration projects where possible.*

Source: PCSD, 1996.

legislation. Pressing problems (for example, climate issues) tend to exceed the ability of further anecdotal responses to deal with them. There is a new realization that global problems cannot be addressed without dealing with all the people's aspirations through 'sustainable development thinking. The President wanted to develop a strategy for the long-term that was integrative and a collaborative effort among different sectors'.

At its first meeting, the PCSD identified four issues it deemed central to considering sustainable development in the USA: sustainable communities, energy, natural resources management, and eco-efficiency. It was further decided to examine three crosscutting issues: population and consumption, education, and tools to stimulate sustainable development. It also focused on goals to enable disagreements to be identified and to see where shared values exist. One PCSD staff member commented, 'We are experimenting in how to get people to say what they think.'

Links to National Planning and Decision-Making

The aim was that the strategy would contain some elements that both the federal government and the private sector could implement readily. There was no existing structure for a strategy to feed into policy-making. But there was a visible desire for change in all the sectors participating in the PCSD process (although some sections of industry and citizens' groups were absent). One PCSD commentator observed:

> *The PCSD will provide a seed bed and a base around which consensus might form. Hopefully, all the major sectors including government will pick up on it. Society is ready for change. But the process of change is not going to be government-led. The chief value of the PCSD is as a channel from society to the government.*

Provincial and Local Strategies

Overall, the PCSD work was not influenced by any regional issues or initiatives. Local-level ones were far more important. Some states have developed innovative environmental and resource programmes that offer models, pilot programmes and demonstration projects for aspects of a national strategy (see Box 20.4) There was early discussion with those responsible for the Minnesota State strategy. The PCSD, especially the Sustainable Communities Task Force, consulted with local and grass-roots organizations involved in sustainable development issues.

Parliamentary Process

It will be for the President to decide whether the strategy developed by the PCSD is debated and pursued, or whether legislation is sought to implement it. It was not designed for Congressional involvement.

Press Coverage

The PCSD sought and received press coverage for all Council meetings. It published a vision, 15 principles, a challenge statement, draft goals and indicators (January 1995), a draft final report (Autumn 1995), and the final report (February 1996). At one Council meeting held in Chattanooga, there was saturation media coverage in the state and modest national coverage. However, since no major controversy seemed to be associated with the work of the PCSD, the press was not particularly interested in covering it during the process. In addition, sustainable development is not a widely understood term in the USA. But once the final report was published and submitted to the

Box 20.4 *SOME STATE INITIATIVES TOWARDS SUSTAINABLE DEVELOPMENT*

Southern California Council on Environment and Development *This is a 'coalition of environmental organizations, citizens' groups, government agencies and private enterprises dedicated to furthering sustainability'. Works via a round-table process to implement recommendations of Agenda 21 in municipalities in southern California and to serve as a forum for promoting sustainable development throughout the region. Other projects include developing sustainability indicators and inventorying current policies, programmes, and costs.*

Governor's Commission for a Sustainable South Florida *This is made up of 35 voting members, including representatives from business, public interest and environmental organizations, and various non-voting government agencies. It is concerned mainly with the Everglades ecosystem.*

Idaho Sustainability Round-Table Information Forum *This is a select group which met once (hosted by the Governor's Office) to address sustainable development issues in Idaho.*

Earth Charter and Agenda 21 for Iowa *This was proposed by the Iowa division of the United Nations Association (UNA), which has been the basis of an ongoing initiative that began in 1991 with a public hearing, community forums and a conference. A report, which has been produced and revised four times, contains three sections — environment and development concerns, education and opportunity, and Iowa youth. The ad hoc working committee for Iowa's Earth Charter includes representatives from the Iowa division of the UNA, the Iowa departments of education and natural resources and various Iowa universities and colleges.*

Kentucky Sustainability Round-Table Information Forum *This met in February 1994, sponsored by the state's Natural Resources and Environmental Protection Cabinet, and its Economic Development Cabinet. It was attended by over 50 participants from the business and labour community, government, public interest groups, civic organizations and academic institutions wishing to learn about round-table methods used in Canada and the USA and to consider how to solve the state's economic and environmental problems. Following a second, smaller meeting, a round table has been established.*

Minnesota Sustainable Development Initiative *Introduced by the Governor in January 1993, this initiative consisted of seven teams, each with 15 members, representing environmental, business, government and public interests. Each team was asked to consider one issue of importance to Minnesota (agriculture, energy, forestry, manufacturing, minerals, recreation, or settlement) and was charged with fact-finding and strategy development tasks. The teams held their final collective meeting on 15 November 1993 to discuss issues of mutual interest and to craft an integrated set of directives and recommendations. They presented a report,* Redefining Progress: Working Toward a Sustainable Future, *at the Congress on Sustainable Development held in February 1994. The report specifies ten principles to guide decision-makers,* *focuses on six central issues and includes a total of 20 strategies. The strategies call for,* inter alia:

- *identifying state laws and policies that send unintentional economic signals that encourage environmentally unsustainable activities while discouraging behaviour the state would like to promote;*
- *redirecting state investments to foster cooperative ventures that promote sustainable development in the public and private sectors;*
- *pursuing opportunities for full cost accounting, including the use of market-based efforts to protect the environment, the development of a comprehensive environmental quality trends monitoring network, and the creation of a standard system of product labelling to inform consumers about the full costs of the products they buy;*
- *fostering sustainable communities through community–state partnerships in strategic planning for sustainable development and growth management; and*
- *use of strategic alliances to make the state a model and to encourage others in Minnesota to work for sustainable development.*

Subsequently, the state legislature adopted a new law calling for a joint legislative-executive task force on sustainable communities and land use. Public meetings were to be held and a Minnesota strategic plan for sustainable development prepared. Momentum was lost in the summer of 1994 when the Governor vetoed a bill that would have provided funding to continue the sustainability efforts of Minnesota.

Missouri: Partnership for Economic Growth and the Environment (PEGE) *This was established in the late 1980s. Participants, including business, environmental and government representatives, work within a cooperative structure of meetings and constructive dialogues to shape/negotiate policy recommendations and develop specific legislative proposals for participants to work on during the next legislative session. PEGE has resulted in specific achievements, including new policies on solid waste and atmospheric protection. In 1991, the participants shaped new legislative recommendations on climate change and ozone depletion. The biggest success has come from the ability of the diverse participants to set aside conventional points of view to reach agreement on several issues each year. A most recent session was planned for early 1995. The PEGE programme is continually expanding and promoting its consensus-building dialogues. The long-term goal is to foster a process throughout the state to achieve sustainable public policies.*

New York, New Jersey and Connecticut: Round Table on the Environment and Economy *In 1994, a group from New Jersey visited the Netherlands to learn from the Dutch experience with the NEPP2 process (see Chapter 14) and 'bought' the idea of consensus-building. A round-table process was established to cover New York and the tri-state area, modelled after the Canadian national and provincial round tables. It involves government, business interests, citizens, environmental groups and others to try to develop a Green Plan. A first meeting was held in March 1993. A steering committee was established, which created a round-table charter and business plan. Topics 'slated' for its agenda include transportation and clean air (New York needs to comply with the Clean Air Act). The steering committee meets four to six times a* *year, while a group of experts meets eight to ten times a year to discuss issue-specific problems facing the environment and the economy.*

Sustainable North Carolina Project *It was initiated in 1991 and coordinated by the Environmental Resource Program (ERP) of the University of North Carolina. It initially reviewed programmes elsewhere on linking economics, environment and equity, and contributed to the development of* Guidelines for State Level Sustainable Development, *published in 1993. ERP then launched a five-year initiative to develop a vision for sustainable development in North Carolina. A 25-person advisory committee was established, consisting of members from various sectors, to help establish structure and vision for the programme. After conducting a state survey, the project's role became that of facilitating participation and consensus among the state's different regional, social and economic groups. A series of regional meetings were planned to raise awareness, establish participants' community values, and set up principles to guide policy development. A state conference was being planned in 1995 to develop policy proposals.*

Washington: (a) Leadership Task Force on Sustainable Economic Development *This was formed by the Department of Community, Trade and Economic Development (CTED) and Forward Washington in August 1993. The task force comprises five working groups and a steering committee, with representatives from government, business, community groups, and environmental and research organizations. Some groups have developed proposals for sustainable development demonstration projects. The initial work scope of the work groups covered energy, natural resources, sustainable communities, public involvement/education, and waste and recovered resources. Briefings on local perspectives on sustainable development have been provided to state and Senate committees.*

(b) Environment 2010 project *As a state project launched in 1988 to develop a clear and comprehensive environmental strategy, it was designed to coordinate the many autonomous agencies and jurisdictions, and to provide mechanisms for anticipating and resolving conflicts. Some 75 recommendations for government, business and communities were developed for each challenge through a combination of public input from citizens attending the Environment 2010 summit in November 1989 and public meetings held throughout the state. Much of this effort is now part of ongoing activities within agencies throughout the state. Two* State Environment Reports *were published (1989 and 1992). Environmental indicators are being developed and a new trends report was due in 1995.*

(c) Sustainable Seattle *This is a volunteer network and civic forum committed to preserving the social, economic and environmental health of the Seattle area. In 1993, Sustainable Seattle published* 1993 Indicators of Sustainable Community, *and the indicators task team was working on an updated version. Other projects include a community outreach project, a policy team that monitors government initiatives, and the marketing and communications team which designs a communications strategy for the sustainability effort.*

(adapted from Nguyen and Roberts, 1994)

President, there was considerable interest, including front-page coverage in the national press and a high demand for copies of the report.

Follow-Up

The PCSD secretariat closed down on 29 March 1996. Implementation will move forward with an interagency working group chaired by Vice-President Gore.

National Environmental Goals Project

Prime Motivation

The Environmental Protection Agency (EPA) initiated the National Environmental Goals Project in 1992 during the Bush administration and it was reaffirmed by Carol Browner when she was appointed Director of the EPA by President Clinton. The project was launched as a response to a 'feeling' among senior EPA staff that while the agency had tried to accomplish various environmental goals in its work, these had not been articulated. It also fulfilled an obligation of the Government Performance and Results Act, which was passed shortly after the project began.

In the past, few EPA staff were concerned with the 'big picture' — most were focused on their own areas of responsibility or interest. But perceptions were changed by an internal EPA report in 1987 ('Unfinished Business') which ranged over the full panoply of issues that EPA was trying to address. A subsequent report of the Science Advisory Body, 'Reducing Risk' (SAB, 1990), confirmed these issues. Both documents opened thinking to broader issues.

Focus

The main focus was to try to change the way people articulate what US public and private efforts in environmental protection should be trying to accomplish. EPA was also concerned with sustainable development targets and pursued these through its participation in the PCSD. There was no intention to reinvent goals where they already existed in government policy, for example, for global climate change. Some 14 sets of targets were being developed (for example, for clean outdoor air, safe drinking water, public awareness and participation). Each of these goal statements (USEPA, 1994) included:

- a problem statement;
- a general goal for condition of environment;
- specific targets for 2005;
- actions to achieve them;
- current status and trends;
- government (federal, state, local) responsibilities; and
- implications for society.

Organization and Management

The project was managed by an EPA goals leadership team (board) comprising office directors and senior regional managers. Staff work on the project was managed by the EPA Office of Policy, Planning and Evaluation (OPPE). A team of about 50–60 EPA staff (from across the country) was engaged in the project.

OPPE staff drafted a first document comprising 13 goals for consideration. These were discussed at round tables, which were organized by regions and states. The round tables were held in nine different cities and attended mainly by people 'knowledgeable' about the environment: federal and state government officials, as well as representatives of business and citizens' groups. Attendance at all round tables and other meetings was by invitation. Meetings were limited by an imperative to maintain a 'manageable size'.

Following further analysis by OPPE staff, the goals were revised in summary form for government-wide policy and technical review in January–February 1995 by the EPA, other federal bodies, states, tribes and the public (see example in Box 20.5) It was intended to complete a version for Congress in the summer of 1995.

Future iterations of EPA's strategic plan 'will focus specifically on the nation's measurable, environmental goals. Meanwhile EPA is developing a goals-based budget for Financial Year 1997 using the current environmental goal areas'.

Terms of Reference

EPA set its own terms of reference for the project, to:

- state clearly the environmental outcomes 'we are seeking';
- engage stakeholders;
- design goal-directed strategies and budgets; and
- report progress towards goal attainment with environmental indicators.

Key 'Assistance' Factors and Problems

Several factors helped the project process. There was a growing realization within the EPA that financial resources were limited and that there was a need to focus activities. Indices (measures) of environmental quality in the USA had slowly been improving. Other countries had also established environmental goals (for example, the Netherlands). The EPA also had to deal with a number of problems. It was necessary to focus on changing behaviour — 'people are used to thinking in terms of activities rather than environmental goals'. When EPA started the project, no very complete set of environmental measurements had been developed. One EPA staff member observed that:

> *Environmental success is not quick to see. Because of 'time lags' in the way the government functions, it is difficult to 'see' environmental improvements and therefore to measure them. For example, when a new set of water quality regulations are introduced, it takes time to develop these, for them to 'permeate' through the system, and for action to take effect.*

Box 20.5 DRAFT ENVIRONMENTAL GOALS FOR CLEAN AIR IN THE USA

GOAL Long-range

Clean Air

The air will be safe to breathe in every city and community and it will be clearer in many areas. Life in some forests and waters will rebound as acid rain diminishes.

Benchmark 1

By 2005, the number of metropolitan areas not meeting air quality standards will be reduced to 6, which corresponds to no more than 45 million Americans living in areas without clean air.

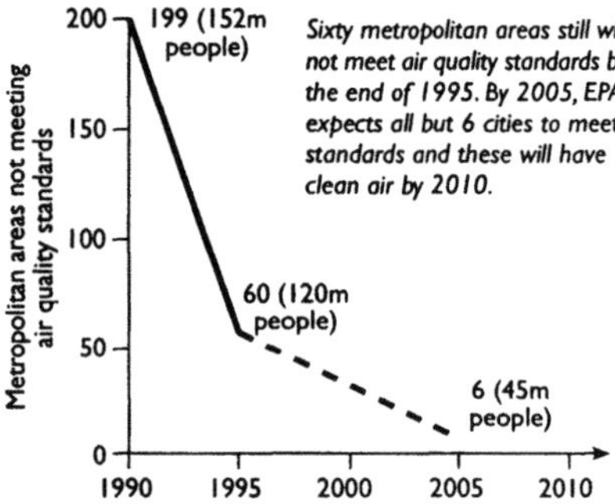

Benchmark 2

By 2005, motor vehicle emissions of smog-causing volatile organic compounds (VOCs) will be reduced 65% from 1990 levels.

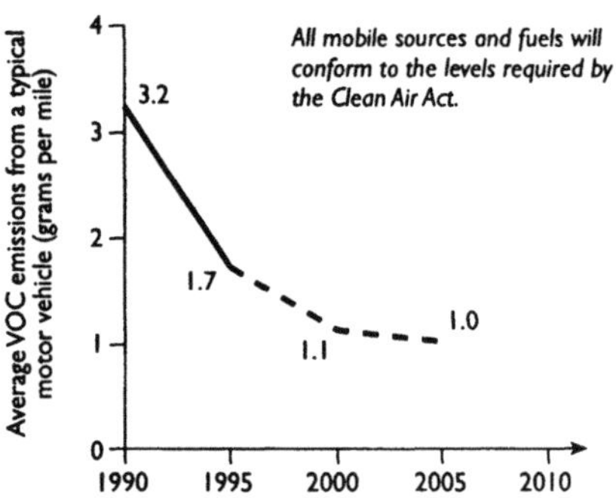

Benchmark 3

By 2005, toxic air emissions from all major industrial facilities will be reduced to the lowest technically-achievable levels.

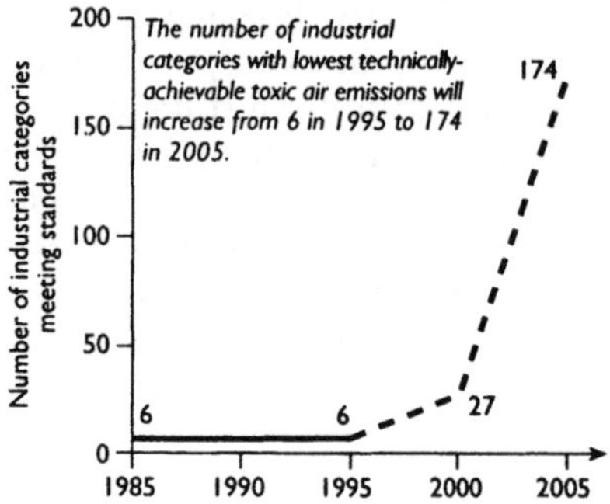

Benchmark 4

By 2005, sulphur dioxide emissions, the primary cause of acid rain, will be reduced 32% from 1994 levels.

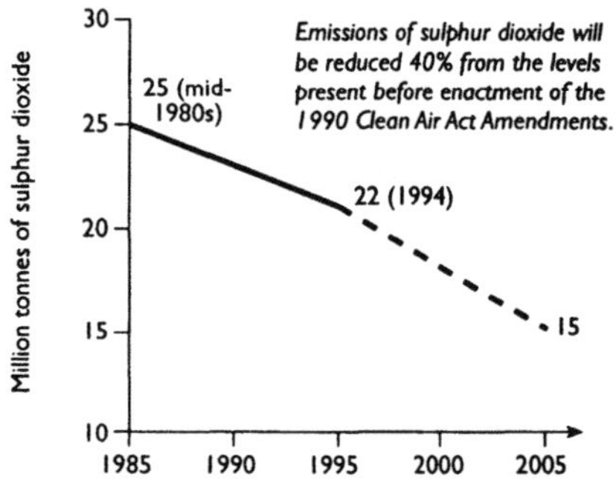

Benchmark 5

By 2005, annual average visibility in the eastern United States will improve from 10% to 30%. The largest improvement (30% increased visual range) will be found in the central Appalachian region.

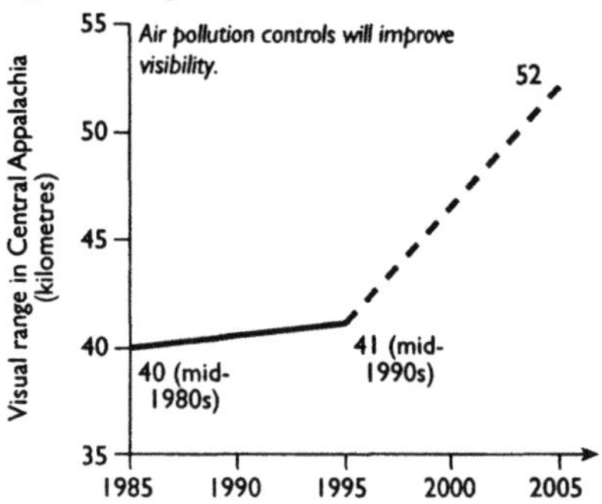

Source: USEPA, 1995.

Chapter 21 THE EUROPEAN UNION

Nick Robins

The main strategy process in the European Union (EU) has been the development of the Fifth Environmental Action Programme (sometimes abbreviated to 5EAP) — *Towards Sustainability: A European Community Programme of Policy and Action in Relation to the Environment and Sustainable Development* (CEC, 1992a). Development of this programme was approved by the Council of Ministers on 1 February 1993. It followed resolutions by the European Parliament on 17 November 1992 and by the Economic and Social committee on 1 July 1992. An interim review of implementation was undertaken by the European Commission (CEC, 1994a), followed by a further progress report (CEC, 10 January 1996a). The European Commission has also made proposals for an action plan (CEC, 24 January, 1996b).

Following UNCED, the Fifth Environmental Action Programme became the European Union's main vehicle for implementing the Rio agreements, and provides a framework for the Community's efforts to implement the climate and biodiversity conventions. Since 1992, the programme's message of environmental integration and sustainable development has been incorporated into a range of European Union strategies and programmes, including the White Paper on growth, competitiveness and employment (CEC, 1993a), the White Papers on transport (CEC, 1993b) and on energy (CEC, 1995), a review of the structural funds, and the fourth research and development programme (CEC, 1994b).

Initiation and Time Perspective

Following the publication of the Brundtland report (WCED, 1987), there was a recognition in the European Commission (EC) and elsewhere in the European Community of the need to articulate a strategy that went beyond traditional environmental protection concerns contained in the first four European Community Environmental Action Programmes and in national approaches.

In 1990, the Irish presidency of the Council of Ministers (representing EC member states) worked closely with the EC to design a new political initiative on the environment. This resulted in the 'environmental imperative' declaration by European Community heads of government at their summit in Dublin in June 1990, which stressed the special responsibility of the EC for the promotion of sustainable development. It highlighted a number of priority objectives and identified key principles (including the right to a clean and healthy environment), which formed part of the Fifth Environmental Action Programme completed at the end of 1991. Other stimuli included the mounting pressure from the European environmental movement, as well as some 'pushing' from a small number of member states for a more strategic approach to European Community environmental policy. With the launch of the Dutch *National Environmental Policy Plan* (see Chapter 14), the EC also had available a model for a new approach to strategy-making for sustainable development. Finally, UNCED was on the horizon.

The Fifth Environmental Action Programme took a longer-term perspective than the EC's previous five-year Environmental Action Programmes. Looking ahead from 1992, it laid down long-term objectives for key issues as 'an indication of the sense of direction or thrust' with a limited number of performance targets for the year 2000. For example, under the theme of climate change, the programme proposed the long-term objective for CO_2 of 'no exceeding of natural absorption capacity of planet earth', with a goal for 2000 of stabilizing emissions at 1990 levels (see Table 21.1).

Focus

The Fifth Environmental Action Programme was presented as a holistic and integrated plan for both the environment and sustainable development. Although there was 'an initial tendency to see it solely as an environmental plan', it is increasingly regarded (within the EU) as a plan for sustainable development as a whole. Its primary role was on the economy-environment interlinkage, building on the Community's primary responsibility for a barrier-free internal market. Unlike Agenda 21, the Fifth EAP does not address explicitly social issues (such as poverty, gender or population), and the European Commission is wary of extending the scope in this direction.

Moving away from the incremental stance of earlier action programmes, the fifth developed a 'new approach' based on the following elements:

- integrating environmental imperative into other policy areas;
- identifying five key target groups responsible for significant environmental impacts — agriculture, energy, industry, tourism and transport;
- pinpointing various priority environmental issues — climate and ozone depletion, acidification and air quality, nature protection and biodiversity, water, urban environment, noise, coastal zones, and waste management;
- changing attitudes and patterns of consumption and production;
- encouraging partnership and shared responsibility;
- recognizing the different roles and responsibilities of different levels of government, from European to local (subsidiarity). (See outline of objectives for municipal waste management in Table 21.2, showing the multiple actors involved);
- broadening the range of instruments (particularly to include economic and voluntary tools);
- improving the application and enforcement of European Community legislation; and

Table 21.1 European Objectives, Targets and Actions for Climate Change

	Objectives	EC Targets up to 2000	Actions	Time-Frame	Sectors
CO_2	No exceeding of natural absorption capacity of planet earth	Stabilization on 1990 level[1] (progressive reductions at the horizons 2005 and 2010[2]	Energy conservation measures, eg environmentally benign energy use; behavioural changes; economic and fiscal measures	Continuous	Energy Transport Industry Public
			Improvement of energy efficiency, for example, R&D; infrastructural changes; change in transport modes; economic and fiscal measures	idem	Energy Waste Transport Industry Consumer
			Fuel substitution towards less or no CO_2 emitting sources (renewables, natural gas), for example R&D; infrastructural changes; economic and fiscal measures	idem	Energy
Methane (CH_4); Nitrous oxide (N_20)		Measures to be identified not later than 1994 and applied (possibly reduction targets)	Inventory of data	Before 1994	Energy Agriculture Waste
CFCs + carbon tetra-chloride + halons + 1.1.1 trichlor-ethane	No emissions of ozone layer-depleting substances	Phase out before 1.1.1996 (except for some essential uses)			Industry
HCFCs, etc		Limitation of use to maximum 5% of 1990 CFC use levels			

Note: The EC commits itself to help and support countries which seek for it, in their aim for stabilization and reduction measures in relation to greenhouse gases. The following measures could be used: debt trading, technology transfer, general trade arrangements, and participation in global fiscal mechanisms.

(1) Targets already set by the EC; (2) Conclusions of the Joint Energy/Environment Council of 29 October 1990.

Source: CEC, 1993a.

Table 21.2 European Municipal Waste Management Objectives, Targets and Actions

	Objectives	EC targets to 2000	Actions	Time frame	Sectors/actors
	Overall target: rational and sustainable use of resources				Industry Agriculture Transport Energy Tourism
Municipal Waste	Prevent waste (closing cycles)	Waste management plans in member states	Landfill directive operational	Before 1995	EC + MS + LAs Industry
	Maximal re-cycling and reuse of material	Stabilize amount of waste generated at EC average of 300 kg/capita (1985 level[1]) on country-by-country basis	Directive on packaging operational	1995	EC + MS + LAs Industry
	Safe disposal of waste that cannot be recycled or reused in ranking order: combustion as fuel; incineration; landfill	Recycling at least 50% of paper, glass and plastics (EU average)	Cleaner technologies product design	progressive	Ind + public + EC + MS + LAs
		Community-wide infrastructure for safe collection, separation and disposal	Policy on priority waste streams, stop landfill for specific wastes (legislation, voluntary agreements)	ongoing	EC + MS + LAs Ind + NGOs + public
		No export outside EU for final disposal Recycling of consumer products	Reliable EC data on waste generated, collected and disposed	1995	EC + MS + LAs + EEA
		Market for recycled materials	System of liability in place	2000	EC + MS
		Reduction of dioxin emissions (90% on 1985 levels by 2005)	Economic incentives/instruments (deposit return systems, voluntary agreements)	ongoing	MS + EC + Ind
			Standards for dioxin emissions from municipal waste incineration	before 1994	EC + MS + LAs

Key: EC = European Commission
EEA = European Environmental Agency
Ind = Industry
LAs = Local Authorities
MS = Member States

Note: Based on Eurostat and OECD statistics.

Source: CEC, 1993a.

- defining the international dimension of European Community sustainability efforts.

The strategy built on the existing principles for environmental policy contained in the Single European Act (for example, polluter pays, and prevention) and anticipated the more systematic focus on sustainable development to be contained in the new Treaty of European Union, signed at Maastricht on 7 February 1992 (which also included the precautionary and subsidiarity principles).

Organization and Management

The Fifth Environmental Action Programme was drafted by a two-person team within the Environmental Directorate General (DG XI) of the European Commission. This team coordinated inputs from various sections of this and other Directorates General. The Fifth EAP document, *Towards Sustainability*, was agreed by the European Commission and put before the Council of Ministers and European Parliament as an EC proposal in March 1992, together with a *State of the Environment in the European Community* report (CEC, 1992b) and a draft political resolution. Following the advice of the European Parliament, the Council of Ministers agreed a resolution which approved 'the general approach and strategy of the programme' in February 1993.

There was no formal process of consultation in preparing the Fifth EAP, which was drawn up before the current 'era of transparency'. Nevertheless, informal consultations took place between the European Commission and NGOs, but these openings were not followed up by NGOs and the EC was criticized afterwards for a lack of consultation. The Fifth EAP 'team' travelled to each of the member states to elicit views and priorities. Once the programme had been approved, the EC funded a series of open meetings in each member state, hosted by the European Environmental Bureau (a Brussels-based organization which represents environmental NGOs from the 15 member states) to discuss its implications.

Reactions

In general, the programme was well received by environmental policy circles within member state governments, businesses, NGOs and local authorities. Over 100,000 copies of the programme have now been distributed in all the EU languages, and this has had a multiplier effect. There is now an awareness of the Fifth EAP within 'specialized circles at the European level', and a general 'greening process has taken place, but it is impossible to judge the causation' (UNCED or other reasons). The European Commission has never pretended that the programme was for 'everyone in the street', and is planning a targeted communications effort for the review of the programme in 1996.

Key 'Assistance' Factors and Problems

The development of the Fifth EAP was assisted by the political impulse provided by the 'Environmental Imperative' declaration by the EC heads of government at the Dublin summit in June 1990, and also by public concern about the environment at the

European level. However, a key problem was the lack of basic understanding among other Directorates General of the objectives of sustainable development, and how these differed from traditional conservation and environmental protection policies. Another difficulty was the problem of developing Europe-wide targets and objectives.

Conflicts and Consensus

During the design and implementation of the programme, disagreements both within the European Commission and within member states have been tackled through 'constant negotiation', with the Environment Directorate aiming to focus debate on key issues so as 'not to waste negotiating capital on lesser problems'.

Links to National Planning

Throughout the development of the Fifth EAP, the European Commission was aware of national planning initiatives (particularly in France — see Chapter 12, in the Netherlands — see Chapter 14, and in the UK — see Chapter 19). The Dutch *National Environmental Policy Plan* (VROM, 1989) had 'an important influence', notably in the target group approach adopted by the EC. Reflecting the principle of subsidiarity, the Fifth EAP does attempt to 'identify where things are done best'. However, the European Commission has never attempted to provide an 'overarching framework' for coordinating European and national-level strategy processes.

Since 1992, the dual impulse of the Fifth EAP and UNCED has led to the publication of national plans for sustainable development in most EU states. The Fifth EAP provided a model for some countries (notably Greece, Italy and Portugal). Other countries were sceptical about the need for a European strategy, but over time a certain value has been acknowledged.

European environmental NGOs, however, continue to press for a common European framework for national strategies, with the European Environmental Bureau proposing that a legally-binding methodology should be devised for securing the 'harmonization and coordination of national plans' and 'citizens' participation' (EEB, 1995).

Links to Local Strategies

The Maastricht Treaty led to the establishment of a new committee of the regions which provides a voice for local and regional authorities in EU policy-making. The committee has established a working group on the environment and provides a major point of contact between the European Commission and local strategy efforts. The EC also supports the sustainable cities campaign, backed by 90 local authorities, which commits its signatories to draw up 'Local Agenda 21s' by 1996. Many local authorities have looked to the Fifth EAP as a model for developing local strategies.

International Dimensions

The Fifth EAP was prepared prior to UNCED and the development of new approaches to North–South relations such as the concepts of 'environmental space' and 'ecological footprints'. However, the programme did recognize that global sustainable development entails 'the equitable distribution and use of resources between nations and regions over the world as a whole'. It also contains a section dedicated to the international dimension.

The mid-term review process has highlighted how the Fifth EAP responds to the Agenda 21 action programme agreed at UNCED (see Table 21.3). The European Commission has also provided financial support for the Sustainable Europe campaign organized by Friends of the Earth Europe — based on the 'environmental space' concept (see Chapter 14), and the General Consultative Forum (established by the Fifth EAP) is developing sustainability scenarios for 2020.

Political and Parliamentary Processes

The special nature of the EU means that political and parliamentary dimensions to the Fifth EAP differ substantially from plans and strategies developed at the national level. The 20 members of the EC (Commissioners) act like ministers in a national government in charge of different policy portfolios. They are appointed by the member states, but take decisions on a collegiate basis. Policies and proposals emanating from the EC are, however, not necessarily linked to particular individuals or parties as in national governments, and the development of policy is a consensual process.

In the case of the Fifth EAP, the EC agreed the programme which the Council of Ministers then approved in general terms, highlighting priority issues, although it had no chance to change the actual programme itself. Prior to this, the European Parliament produced an 'advisory opinion' which made a series of non-binding requests to both the EC and the Council of Ministers, some of which were taken on board.

Following the adoption of the Maastricht Treaty, the European Parliament's role in the review of the Fifth EAP will be much stronger. Under the Treaty of European Union, multi-annual programmes, such as the Fifth EAP, have to be approved by the 'co-decision' procedure. This strengthens the bargaining position of the Parliament by giving it an effective 'veto' of proposals, and forcing the Council of Ministers to go to conciliation if it does not accept the Parliament's amendments at a second reading. The approval of the action plan to update the Fifth EAP will therefore involve the negotiation of compromises between the EC, the Council of Ministers, and the European Parliament. This could result in a strengthening of the text, given the pro-environmental sentiments of the Parliament.

Implementation Mechanisms for Follow-Up

(a) Within the Commission

- all new actions by the Commission have to be examined from the environmental point of view;
- new legislative proposals with a significant impact on the environment must both describe and justify the impacts;
- each Directorate General has appointed a high-level official closely linked with policy-making (an 'integration correspondent') to ensure that policy and legislative proposals take account of the environment and contribute to sustainability;

Table 21.3 Extract of Comparison of Agenda 21 and the European Fifth Environmental Action Programme

Agenda 21 (adopted June 1992)	Fifth EAP (adopted by EC on 18 March 1992, endorsed by European Parliament and Council of Ministers end of 1992)
Coverage UN and its member countries World and North–South	*Coverage* EU and its member states European policies
Statute non-binding framework: political commitment at the highest level up to 21st century, which each should apply at its own level and within its own means	*Statute* framework of environmental and other policies for the period 1993–2000, applicable to European and national policies, in line with subsidiarity
Structure 700-page document divided into four sections, each divided into chapters: – horizontal aspects – themes – principal groups – methods	*Structure* 100-page document divided into three parts, each divided into chapters: Part I: the Community Part II: the wider international arena Part III: follow up Annex: State of the Environment Report
chapters contain: – principles for action – objectives – activities – means of implementation including financial cost	chapters contain: – long-term objectives – objectives up to the year 2000 – specific actions – actors
Section I: economic and social dimensions: trade, poverty, consumption patterns, demography, human settlements, health, integration in the decision-making process	Part I: included in Chapter 5: themes Chapter 7: broadening the range of instruments

Source: CEC, 1994a.

- the Environment Directorate General (DGXI) has set up its own unit to monitor and coordinate implementation;
- each Directorate General is responsible for undertaking a regular evaluation of its environmental performance and for preparing a report on its progress;
- a code of conduct is in preparation for internal Commission operations (such as purchasing, transport and maintaining the efficiency of buildings); and
- public reporting on progress in integration is made through the Commission's annual report.

These measures have now been incorporated into the *Manual of Procedures* issued to each official. According to a progress report published in January 1996, the 'measures so far have had a limited impact'.

The establishment of the network of 'integration correspondents' throughout the Commission has meant that there are now senior officials in each policy area 'fighting in the same direction'. Thereafter, however, there are still 'great variations' in the degree to which the Fifth EAP has been internalized.

For the Environment Directorate General (DGXI), the priority has moved on from explaining *why* integration is need to showing *how* it can be done. This requires a new approach within the Environment Directorate General, forcing it to think in a more 'service-oriented' way towards other Directorate Generals, and requiring a new set of skills from environmental officials. Training programmes on negotiating skills are being conducted.

(b) Within Member States and Others The Fifth EAP led to the establishment of three *ad hoc* dialogue groups to help realize the message of shared responsibility, including:

- the General Consultative Forum, made up of 32 individuals from business, labour, environment, and local government, appointed in their personal capacities rather than on a representative basis, to advise on the best approach to sustainable development at the European level. One major output of the forum has been a statement of 12 principles on sustainable development;
- the Environmental Policy Review Group, bringing together DGXI and the environment ministries of member states to monitor the implementation of the Fifth EAP. These meetings have become important occasions for reflection and sharing strategic information; and
- an informal network on implementation and enforcement of legislation (IMPEL), which has focused on the exchange of information and experience.

(c) The Midterm Review, 1995–6 The Fifth EAP included a commitment to a midterm review in 1995. The review actually started in 1994 with the publication by the EC of an interim report in November 1994, assessing the progress made in the first 18 months. This report concluded with a message of 'cautious optimism', stating that

> *the combination of the Fifth Programme and the Rio Earth Summit, underpinned by the legal and procedural changes incorporated in the Treaty of European Union has marked a turning point in the way sustainable development is being approached both in the Commission and in Member States. [Despite this, the report also argued that] sustainable development essentially continues to be seen as the business of those who deal with the environment.*
>
> (CEC, 1994a)

The European Commission began the full review in 1995. There was extensive consultation, with a 'conscious attempt to give other people a greater sense of ownership'. A number of routes were used to gather inputs:

- the EC sent out 200 letters to a wide range of European interest groups, requesting their input;
- the EC requested input from the environment ministries of member states, together with NGOs, businesses and local authorities. Most member states sent considerable amounts of written material; some sent nothing. The EC also held discussions with national administrations, some of which involved the whole range of interested organizations in the discussions;
- the Environmental Policy Review Group and the General Consultative Forum were involved in the review process and provided suggestions; and
- the European Environmental Bureau (representing NGOs) held meetings in each member state to decide on priorities for the review, and published a comprehensive protocol for turning the programme 'from a strategy to a plan' (EEB, 1995).

The first draft of the review was sent to the member states for comment, and workshops were held for five target groups (agriculture, energy, industry, tourism and transport), NGOs, local authorities, and other bodies, to check the findings. A large degree of consensus emerged, except in the case of transport — now seen as one of the most contested environmental issues facing the EU.

The final progress report was published by the EC in January 1996. This included a comparison of Agenda 21 and the Fifth Environmental Action Programme (see Table 21.3). It concluded that 'the overall strategy and objectives of the Fifth EAP remain valid; what is lacking are the attitude changes and the will to make the quantum leap to make the necessary progress to move towards sustainability' (CEC, 1996).

This confirmed the assessment reached by the European Environment Agency in November 1995 in an updated state of the environment report that 'if the European Union wants to achieve its environmental targets, an accelerated environmental policy is needed'. In the spirit of 'shared responsibility', the progress report covers efforts to move towards sustainability at the European, national and local levels.

Follow-Up: A New Action Plan

On 24 January 1996, the European Commission adopted an action plan with priority areas and actions at the EU level based on findings of the progress report and the report of the European Environment Agency. The overall approach and strategy of the programme remains valid, but priority setting and more operational action are needed to give a new impulse to sustainable development and the implementation of the Fifth EAP. The following key priorities are identified:

- actions are needed to integrate the environment into the five main policy sectors: industry (with focus on small and medium-sized enterprises, SMEs), agriculture, energy, transport and tourism;
- a broader mix of policy instruments, particularly focusing on economic and fiscal instruments, are required;
- legislation is still needed for part of the policy instrument mix and simplification; as are better implementation and enforcement of environmental legislation;
- increased efforts need to be made to raise the awareness of European citizens in order to bring about changes in behaviour; and

- the EU needs to assume more leadership on international environmental issues.

It is noted in the plan that success in these areas will only be possible if actions are supported by better data on the state of the environment, and if environmental and performance indicators, sound scientific information, and improved cost-benefit approaches are forthcoming.

The action plan has been sent to the Council of Ministers and the European Parliament for agreement by the new co-decision procedure as laid down by the Maastricht Treaty. This could ensure a greater degree of political commitment to achieving EU environmental objectives than was possible with previous programmes.

Independent Initiatives

Following development of the *Action Plan: Sustainable Netherlands* by the Dutch Friends of the Earth (Milieudefensie, 1992), based on the concept of environmental space, Friends of the Earth Europe launched the Sustainable Europe project to encourage debate and to produce a report providing a European perspective and a vision on sustainable development (see Box 21.1) (see also Wuppertal Institute, 1995).

Box 21.1 SUSTAINABLE EUROPE PROJECT

The Sustainable Europe project aims to encourage debate and to produce a report providing a European perspective and a vision on sustainable development. The definition of sustainability used includes a limited set of quantified targets, based on the concept of environmental space. The project has been coordinated by Milieudefensie (partly sponsored by the International Affairs Department of the Dutch Ministry of Housing, Spatial Planning and the Environment and by the EC–DG11). Among other issues, this project deals with transboundary questions within Europe. In 1993, the Wuppertal Institute for Climate, Environment and Energy in Germany was contracted to define 'environmental space' for Europe and to develop a framework for individual European country studies. This led to the preparation of a draft handbook, so that the discussions being organized with the various actors in each of some 30 participating countries would have the benefit of these studies.

A workshop was held in April 1995 in Brussels involving all FoE participants to prepare for a conference in November 1995 involving major actors from all over Europe. Some 140 participants from 32 European countries (representing business and industry, governments, trade unions, consumer organizations, research institutions, political parties, government institutes, as well as environmental and nature protection groups) discussed in detail the necessary steps to bring Europe closer to sustainability. The discussion led to six recommendations with explanatory texts:

(1) *National and local authorities and international bodies (as appropriate) should, after wide dialogue and negotiation with all relevant social actors, including businesses, NGOs and communities, and drawing on the best scientific evidence, set, and as necessary revise, national sustainability targets for levels of consumption of environmental resources based on the concept of global environmental space. They should also develop visions and strategies to strive to reduce significantly consumption of key resources such as energy and raw materials by 2010, and to reduce consumption to within the available environmental space in the following decades.*
(2) *National authorities and international bodies like the EU should, after wide dialogue with all relevant social actors, adopt and regularly publish indicators of sustainable economic welfare, incorporating measures of life and environmental quality, and use them to help guide fiscal, economic and social policy, and integrate such policy with environmental goals. Programmes for economic support within and between countries should seek to increase the quality of life and human wellbeing.*
(3) *National governments should work, individually and together, to introduce a revenue-neutral 'ecological' tax reform, preferably at a pan-European or EU level, increasing taxes on environmental resources (with appropriate compensation if necessary for the least well-off), using the revenue to reduce taxes on labour and to create new, environmentally sound job opportunities. Supporting regulation, planning processes, public expenditure and other economic instruments should be*

designed to drive the development, and encourage the uptake, of environmentally efficient technologies.

(4) *We must seek an equitable distribution of access to environmental space between North, South and East. All policies and operations of governments and international development agencies, including approaches to trade, aid, international debt and technology transfer, should take full account of quantified sustainability targets. The present Western development paradigm is unsustainable, and should not be promoted in the South and East. Eastern countries should take the opportunity provided by the transition period to move directly to sustainability. Global cooperation and mutual learning are crucial.*

(5) *Businesses should take voluntary action, with a long-term perspective, above and beyond regulation, to develop cleaner technologies, products and management practices and transmit them throughout the world, with the aim of meeting consumer needs within sustainable lifestyles.*

(6) *Local, national and international authorities and communities must seek to empower all individuals and NGOs of all kinds to take a full role in local, national and international decision-making processes. Human rights, particularly the right to know and the right to participate in public hearings, must be supported and implemented by governments and businesses. Such rights are accompanied by responsibilities for enhancing sustainable personal behaviour. Public environmental awareness and commitment to sustainable lifestyles must be increased through education, training and communication. NGOs, educational bodies, churches and other relevant actors should take an active part in all these processes.*

Source: FoE Europe, 1995.

Chapter 22 GREEN PLANNING and STRATEGY PROCESSES in Some Other INDUSTRIAL COUNTRIES

Austria

In 1992, the Austrian government initiated the development of a *National Environmental Plan* (*National Umwelt Plan*, NUP), which would seek to integrate environmental quality objectives into all policy sectors and outline Austria's long-term environmental management strategy for sustainable development. The plan, adopted by the government in May 1995, 'is meant to serve as a binding frame of reference for all concerned'. The document represents the end of a 'development phase', based on a 'broad consensus in the fields of science, economy and politics'. An English translation is available (AustMoE, 1995). Seven working groups were set up, consisting of government officials and representatives of industry, trading and agricultural associations, employers' organizations, environmental NGOs and the scientific community. They were asked:

> *to operationalize and implement the principle of sustainability in environment-related policy sectors in Austria. A key objective was to define long-range and strategically oriented ecological goals of a qualitative and quantitative nature; the vision was to transcend traditional media and sectoral boundaries. Furthermore, integrated medium- and long-term concepts of environmental care were to be developed and firmly anchored in the body politic.*
>
> (AustMoE, 1995)

The new NUP provides Austria with a long-range concept that operationalizes the political commitment to integrate environmental concerns into all political levels; this includes policies for industry, traffic and energy, agriculture, health, research and technology, as well as education. The plan treats a range of overlapping socioeconomic and ecological fields such as the relationship between society, community and the individual, problems related to resource management, consumption patterns and consumer behaviour, as well as sectoral issues (as listed above).

Building on national, European and international norms, measures and experience, the NUP's task was to define long-term objectives and standards for Austria. A catalogue of the measures necessary to promote 'environmentally compatible development', including the underlying structural changes, was developed. The key elements of the plan include long-range quality targets based on scientific criteria, as well as proposed measures to reduce pollutants, to make sensible use of non-renewable resources, and to minimize material flows.

The task of the NUP is stated to be an 'open, innovative, evolutionary process whose future configuration will be amended by a steady flow of new knowledge, data, understanding of interdependencies, and probably also a new perception of environmentally sound development. The NUP will therefore be evaluated periodically'. Implementation is first to be assessed after two years and the plan will be revised and updated every four years.

A youth environment plan was set up as a complementary initiative to allow Austria's youth to participate in national environmental policy. It is a link between the NUP, the ministries responsible for the environment and for youth, and those schoolchildren, apprentices and students between the ages of 15 and 25 who are interested in environmental issues. This is a unique model in Europe. It provides a forum for the youth to ventilate their views and concerns about the goals of future environmental policy.

Germany

The *German Strategy for Sustainable Development* (GMoE, 1994) outlines Germany's environmental policy for the 1990s. This is driven by the desire to integrate economic, social and environmental concerns.

The initial section of the strategy report reviews the policy challenges posed by Germany's recent political reunification, opportunities for environmental cooperation within the European Union and internationally, and, in particular, the responsibilities of the new Laender (former East Germany) to resolve the legacy of environmental degradation by developing environmental infrastructure, institutions and legislation.

The strategy document also reviews:

- specific policies governing chemical and waste management, air and water quality, natural resource conservation (including soil and landscapes), forest management, noise control, and nuclear energy;
- Germany's intention to incorporate all environmental legislation into the basic law; and
- current efforts to incorporate environmental protection efforts into other areas of action and policy, including energy use, traffic management, agricultural activities, settlement development and land-use planning, leisure activities, and defence.

The concluding sections describe the impetus behind Germany's environmental research activities, and its international environmental cooperation with central and eastern Europe, the new independent states, and Western industrialized countries.

The Federal Ministry for the Environment, Nature Conservation and Nuclear Safety and the Federal Ministry of Economic Cooperation and Development are jointly responsible for the coordination of follow-up to UNCED in Germany, particularly the implementation of Agenda 21.

The activities of NGOs in UNCED follow-up are coordinated by the German NGO Forum on Environment and Development. An important contribution was a study commissioned jointly by two NGOs (Bund für Umwelt und Naturschutz Deutchland, and Misereor) from the Wuppertal Institute for Climate, Environment and Energy, *A Viable Germany: A Contribution to Global Sustainable Development* (Wuppertal Institute, 1995). This study addresses the issues related to life in a future Germany that is viable in terms of the environmental aspects of development.

In October 1995, Berlin became one of the first German cities to produce a draft of a Local Agenda 21. In 13 chapters, this summarizes the main measures and results of Berlin's environmental policy, and defines objectives for the coming years. The German government is providing financial support for the elaboration and implementation of a Local Agenda 21 for the Köpenick district of Berlin.

Ireland

The *Environmental Action Programme* (IDoE, 1990) set out objectives for protecting and improving the environment in Ireland across a range of public sector activities. Targets involving public expenditure of almost £1 billion over a ten-year period were identified in relation to a variety of measures, including pollution control, preservation of the natural and built environment and the enhancement of amenities. The programme indicated investment of £930 million, in particular for the provision of new and improved public water supplies and sewage treatment services. Measures concerning waste (particularly promoting recycling), agriculture, forestry, energy and industry were also included. The programme proposed the creation of an Environmental Protection Agency (subsequently established in July 1993) to exercise environmental management over all activities with major polluting potential and to provide important supervisory and support services to other public authorities, including local authorities.

In June 1995, the Irish Department of the Environment published *Moving Towards Sustainability: A Review of Recent Environmental Policy and Papers* (IDoE, 1995a), which points out that Ireland has made progress in recent years with environmental protection, incorporating more fully into environmental legislation considerations of sustainability, precaution, cost allocation and integration. But problems clearly remain:

> *Waste generation and energy consumption have been increasing, as has the incidence of slight to moderate water pollution in Irish rivers and lakes. Economic growth, which is now establishing itself in Ireland, cannot be permitted simply to add to these pressures. On the contrary, sustainable economic development must seek to break the link between increased production and greater consumption of environmental resources.*
>
> (IDoE, 1995a)

The report outlines recent policy developments and provides an overview of recent measures to increase environmental protection. It identifies the systems that have been established and the actions that have so far been taken to promote sustainable development in Ireland. In this way, it provides a background context and a point of reference for the preparation of the national sustainable development strategy. The latter is being taken forward by a Cabinet subcommittee and supported by a range of other initiatives, including:

- The definition of a new set of sustainable economic development indicators to be used alongside GDP. The National Economic and Social Research Institute has been commissioned to advise on the formulation of such indicators for Ireland. At the start of this work, a special forum was held in April 1995, organized by the Universities Research Group on the Environment and *An Taisce* (the National Trust for Ireland), supported by the Department of the Environment. This brought together people with a wide range of interests to discuss issues related to the development of sustainability indicators.
- The formulation of environmental management plans to give public sector leadership in demonstrating best environmental management practice.

The national sustainable development strategy will be completed in 1996. It will address all areas of government policy that impact on the environment. Its implementation process will involve the targeting and measurement of key objectives, as well as annual reviews (IDoE, 1995a).

While work is being led by the Minister for the Environment and the Department of the Environment, there are a number of strands involved in its development. Within government, a Cabinet subcommittee will agree the final shape of the strategy; at official level, sectoral objectives and targets are being considered and refined by the Green Network of Government Departments (set up in 1994 to promote better exchange of information between the departments concerned on issues of environmental importance and to support the work of the environment units established in the main economic departments). Within Parliament, a joint committee of both houses of the *Oireachtas* (Parliament) — the Joint Committee on Sustainable Development — has been established to identify, in anticipation of the strategy, measures to maximize the advantage of environmentally sustainable sectoral policies. Local government and local communities have been brought into the process through the publication of guidelines on Local Agenda 21 (IDoE, 1995b), and a national conference on Local Agenda 21 was held in June 1995.

A final consultation was being arranged to enable interested bodies and organizations, or individuals, to make written submissions for consideration in the context of the strategy preparation. A public notice to this effect was placed in the national press in early 1996.

Japan

In November 1993, Japan enacted the Basic Environment Law which lays down basic principles and policy direction, and provides the basis upon which to implement a comprehensive environmental policy. The BEP (Basic Environment Plan) is the most important measure introduced under the new law. Approved by the Cabinet in December 1994, the plan clarifies the long-term objectives of environmental policy, looking ahead to the middle of the twenty-first century, and specifying measures to be taken by the government up to the early twenty-first century. There is an English translation (EAJ, 1994) and an English outline (EAJ, 1995).

The BEP sets four long-term objectives: building a socioeconomic system fostering a sound material cycle; harmonious coexistence between humankind and nature; participation by all sectors of society; and the promotion of international activities. It spells out not only policies taken by the government but also the roles and activities expected of local governments, corporations, the people and private organizations for achieving those objectives. Furthermore, it provides mechanisms for implementing these measures effectively.

To promote measures aimed at particular environmental issues, existing goals in specific areas will be revised as necessary, following the general direction of the BEP. Furthermore, concrete targets and specific programmes will be developed as necessary in areas where they do not exist. The government has indicated that it will adopt appropriate financial and other procedures required to implement the measures contained in the plan, which is also seen as the country's basic formula for environmental conservation.

> *It is important that between this Plan and other national plans harmony with the Basic Environment Plan should be secured on matters related to environmental conservation. As such, we will make careful coordination in line with this Plan to realize its aims in various other national plans.*
>
> (EAJ, 1995)

Indicators are also being developed to measure the actual progress made toward the four long-term objectives. These indicators will be used to check on the implementation of the plan every year, and the plan itself will be reviewed after about five years, so that it can respond to changes in society and the economy.

As a follow-up to UNCED, Japan has also prepared a *National Action Plan for Agenda 21* (Government of Japan, 1993). This document discusses many of the issues raised by Agenda 21 (UNCED, 1992) in detail in relation to the Japanese situation. It contains 40 chapters in four sections, covering social and economic dimensions, conservation and management of resources for development, strengthening the role of major groups, and means of implementation.

In its preamble, the action plan attaches importance to the implementation of the following measures:

- making efforts to construct a society in which sustainable development imposes a reduced load on the global environment, and to enhance public awareness with a view to making people's lifestyles more environmentally-friendly;
- actively participating in, and contributing to, the creation of an effective international framework with regard to the conservation of the global environment;
- actively taking part in the international negotiations for setting up effective financial mechanisms, particularly the restructuring of the Global Environmental Facility (GEF), in order to deal with the conservation of the global environment;

- making efforts to promote environment-related technological development, and contributing to the capacity-building of developing countries to address environmental problems through providing appropriate and well-planned official development assistance, including the promotion of technology transfer;
- ensuring international cooperation for observation and surveillance, and surveys and research that relate to the conservation of the global environment, as well as actually carrying out these activities; and
- enhancing the level of effective cooperation among major constituents of society, including the central government, local authorities, businesses and NGOs.

The plan posits itself as 'an important first step for Japan in realizing the conservation of the global environment through sustainable development'.

Portugal

Portugal has recently been engaged in a period of intense planning activity in conjunction with the preparation of its regional development plan for 1994–9, as well as the subsequent approval of a new community support framework. In November 1994, the *National Environmental Plan* was submitted to the Council of Ministers and was approved on 21 April 1995 (by Resolution No 38/95). An English language summary of the final plan is available (IPAMB, 1995). According to this summary (p2), the plan:

> *should be viewed as an instrument that endorses the consistency of the investments that will be carried out in the specific area of the environment, in order to strengthen the relationship between the environment and other areas of government, within the framework of overall development of the Portuguese society, and above all, it should establish an environmentally sound policy that, from now on, may be projected beyond the current investment cycle, so that it may increasingly and irreversibly lead the Portuguese society to a model of sustainable development.*

Following the introduction, the *National Environmental Plan* is structured as follows:

Chapter 2 provides details of the Strategic Guidelines of the Environmental policy.

Chapter 3 presents the objectives and specific actions of the Ministry of the Environment and Natural Resources (MARN), i e a set of fields of action and the chief measures taken within their scope, where the services of MARN have specific responsibilities or are specially relevant to their action. The topics contemplated are water resources, the coastal area, nature conservation, urban and industrial wastes, air quality and climate, noise, the urban environment, natural risks and those generated by man, consumer protection, research and technological development, education, training and public participation, and institutional capacity-building.

Chapter 4 deals with the objectives and cross-sectoral programme actions, addressing in a systematized way the different areas of government. Special emphasis is given to the areas seen as target sectors within the scope of the Fifth European Action Programme (see Chapter 21), in particular industry, energy, transport, agriculture and tourism, as well as the sea and planning, due to their direct relation with the environment and land management.

Chapter 5 systematically addresses a set of instruments for the implementation of the *National Environmental Plan*, seen as horizontal fields of action that are relevant at all levels of environmental concern. Some of the instruments are often referred to in Chapters 3 and 4, when the specific fields of action are mentioned, but their importance and structural nature are highlighted.

Chapter 6 proposes a system of evaluation and monitoring of the *National Environmental Plan*. This is essential to assure its practical implementation, to prevent it from becoming a mere speculative exercise, and to grant continuity to the attention given to it by individuals and institutions during its preparation and public discussion.

Finally, eight annexes are included, which provide brief descriptions of the state of Portugal's environment; the occupational and organizational structure of the country; the status of Portuguese public opinion on the environment; the sources of financial support for the plan; the availability of instruments for land planning and management; the main powers of the central and local public administrative bodies *vis-à-vis* environmental matters; the global evaluation criteria used for environmental projects; and environmental planning and management indicators.

Russia

A request to the government to prepare a document setting out a 'Concept of Russia's Transition to Sustainable Development' was made by President Yeltsin in a decree (executive order) in February 1994. The government issued an order in March 1995 in which it placed the Ministry of Economy in charge of an interagency committee to undertake the task. The committee was chaired by a deputy Minister of Economy and the vice-chairman was the deputy Minister of the Environment.

The Ministry of Economy established an expert working group to prepare a text on the concept. The first draft, completed in late 1994, was sent to all federal ministries and agencies, and the administrations of Russian provinces within the Russian Federation. It was published in the Russian environmental weekly, *Zeleny mir* (Green World), in February 1995. Comments and critical remarks were invited and taken into account in preparing a second draft, which was subsequently discussed by the Russian Congress on Environmental Protection in early June 1995. The Congress recommended that work on the sustainable development concept should continue and a new and expanded drafting group was set up for the purpose.

The final draft was completed in November 1996 and was submitted by the Ministry of Economy to the government, which considered it at its session in late January, chaired by Prime Minister Victor Chernomyrdin. The text was approved with some recommended changes. The government secretariat assumed responsibility for the finished document.

On 1 April 1996, President Boris Yeltsin issued Decree No 440 'On a Concept of Russia's Transition to Sustainable Development', which adopted the text of the concept (which was attached to the decree) and set out the direction of two further major activities, namely that the main provisions of the concept should be taken into account in new legislation and in economic and social development decisions; and that a strategy for Russia's transition to sustainable development should be prepared in 1996.

The concept is based on the triad of ecology, economy and society. It spells out goals for ecological integrity, particularly those of observing the carrying capacity of ecosystems, eco-efficiency and eco-equity. Broad criteria for achieving major objectives are also indicated. Special chapters are devoted to regional sustainable development issues and their international dimension.

The Presidential Decree was followed by a government ordinance, signed by the Prime Minister on 6 May 1966, placing the Ministry of Economy in charge of elaborating a national strategy for sustainable development in Russia together with the Ministry of Environment, and the Ministry of Science and Technological Policy. Other ministries, regions, NGOs and scientists will be enlisted to do the work. Three deadlines have been set. By 1 June, guidelines should be prepared for designing a sustainable development strategy in line with UNCED recommendations. By 1 September, regions should submit their proposals. By 1 November, a draft sustainable development strategy should be made and submitted to government. The work on the strategy should take into account, in particular, the concept of Russia's transition to sustainable development, the programme of Russia's socioeconomic development for 1996–2000, and the forecast and concept of Russia's socioeconomic development for 1996–2005.

Appendix 1
Questionnaire for Structured Interviews

A: How did the strategy get going and why?

1. What strategy (or near equivalent) processes (a) have been undertaken, (b) are underway or (c) are being planned in your country?
2. Which one(s) have you been involved in and in what capacity?
3. When was this (or were these) initiated and by which institution (for example cabinet, line ministry, NGO)? What is its/their time perspective (for example, 3 yrs, 20 yrs)?
4. In respect of question 1, has more than one national strategy in the country overlapped in time, scope and content? Have past or existing strategies been displaced, undermined or duplicated? What has been the result of this (for example, confusion, wasted money or professional inputs)?

B: What were/are the main aims and the focus?

5. What was the prime motivation for it/them (for example, UNCED, Agenda 21, commitment under international convention, response to public pressure, independent decision of government, cabinet, ministry)? — please be frank!
6. Was/is there a set (or even hidden) purpose or aim of the strategy?
7. What was/is the main focus of the strategy (for example, conservation, physical planning, environmental action planning, more holistic and integrated planning for sustainable development)?

C: How was/is the strategy being organized and managed?

8. Who was/is responsible for the strategy process (eg, a government agency, NGO, independent secretariat, etc)?
9. What structures and strategy management systems were/have been established?
10. What were/are the terms of reference and who set these?

11. How long did/will the process take?

D: Who participated in the strategy?

12. Who was/is involved and how (nationally, regionally, locally, government, NGOs, private sector)?
13. How was the process and/or the product (i e the strategy document) received by NGOs?
14. To what extent was/is the process:
    (a) consultative (for example, organizations and people were/are being asked to comment on documents or proposals)?
    (b) participative (organizations and people were/are involved in determining the process itself and making inputs to the development of the strategy?
    Who was/is involved in consultation/participation?
15. Was an official mandate set for the strategy? Was the process mechanism or duration predetermined by higher authority within government (i e so that you were obliged to follow a particular approach)? Were there alternatives which were excluded and why? If so, what effect did this have on the strategy?

E: What were/are the key factors, issues and problems?

16. Were there any factors that particularly assisted the development of the strategy (for example, a past strategy as a point of departure, existing public pressure and willingness to cooperate, genuine existing government commitment to pursue a route towards sustainable development)?
17. What were the key problems faced during the development and implementation of the strategy?
18. Were there any key issues that the strategy had to resolve (for example, environmental degradation, land ownership, poverty) which determined the approach/process?

F: How were/are problems and conflicts solved?

19. Were there any difficulties in defining or resolving different opinions about these issues? How were such difficulties resolved? How were choices made? What processes did you use to resolve conflicts/disputes: arbitrators, drafting sessions, round tables leading to compromise texts?
20. To what extent is there consensus about the process and content of the strategy (a) across government, and (b) across society?

G: How did/does the strategy relate with other strategies?

21. How did the strategy process link to existing national/local planning systems and decision-making systems in the country?
22. Are there any relevant regional (i e European, North American) strategies in existence or being developed? How did the national strategy draw from, input to or relate to these regional strategies?

23. Are there any local (for example, district- or city-level) strategies in existence or being developed in the country?
    What is the relationship between these and the national strategy?
24. How will the requirement to produce convention-related national action plans be dealt with in relation to the national sustainable development strategy (or near equivalent) process? [The recently concluded conventions on biodiversity, climate change and desertification each require signatory countries to develop national action plans.]
    Where your country has signed such conventions, what are the administrative, institutional and financial implications of undertaking so many plans?
25. Does your country have responsibility for any dependent territories? If so, does the national strategy cover these territories and how were they involved?

H: What was/is the driving perspective and were/are wider issues dealt with (for example, ecological footprints, transboundary issues)?

26. Has the strategy process and decision-making on recommended actions been driven by a particular perspective (for example, environmental) or a central discipline (for example, economics) or has there been an interdisciplinary approach?
27. Does the strategy address:
    (a) the issue of the nation's 'ecological footprint' (i e the impact/effect of the nation(s) activities — such as trade — on the environment in the wider world); or 'environmental space (i e available per capita global carrying capacity for particular resources)?
    (b) cross-boundary issues (particularly with respect to neighbouring countries); and were there any cross-boundary processes to discuss negotiation and compensation?

I: Has the strategy led to parliamentary and wider debate?

28. Was there or will there be a parliamentary process concerning the strategy, for example a parliamentary committee or debate on the strategy and the issues it raises?
29. To what extent and how has the national strategy facilitated a 'greening' of the political, business and consumer mainstreams, and of the values, lifestyles and choices that underlie and shape them?
30. Did/has the strategy received any national press coverage? Was it extensive? Can you provide copies of such coverage?

Appendix 2
Useful Contacts

The following people who were interviewed and/or consulted during this study may be able to provide more information about individual strategies or green planning initiatives discussed in this book.

Australia

National Strategy for Ecologically Sustainable Development
Paul Garrett
Director, Intergovernmental Unit
Coordination and Liaison Branch
Environment Strategies Directorate
Department of Environment, Sport and the Territories (DEST)
GPO Box 787, Fifth Floor, NCC House
16 Moore Street, Canberra City, Canberra, ACT 2601
Tel: +61–6–274–1387
Fax: +61–6–274–1858

Canada

Projet de société
Executive Director
National Round Table on the Environment and the Economy
1 Nicholas Street, Suite 1500, Ottawa, Canada
Tel: +1–613–992–7189
Fax: +1–613–992–7385

Green Plan
Richard Smith
(Director-General, Planning and Coordination, Policy and Communications)
Wilma Vreeswijk
(Director, Planning and Priorities, Policy and Communications)

Stephen Blight (Policy Manager)
Environment Canada
22e Les Terraces de la Chaudière, étage 10
Rue Wellington
Hull, K1A OH3, Quebec, Canada
Tel: +1–819–994–1670
Fax: +1–819–997–0709

Denmark

Nature and Environment Policy
Marianne Rønnebæk
Ministry of the Environment
Højbro Plads 4, DK–1200, Copenhagen K, Denmark
Tel: +45–33–927600
Fax: +45–33–322227

European Union

Fifth Environmental Action Plan
Robert Hull, and Robert Donkers
Coordination and Integration Unit
Environment Directorate-General
European Commission, Brussels, Belgium
Tel: +32–2–299–2289
Fax: +32–2–299–0895

France

Plan national pour l'Environnement
Pierre Guelman
Chef du Bureau de la Coordination interministérielle
Direction générale de l'Administration et du Développement
Ministère de l'Environnement
20 Avenue de Ségur, 75302 Paris 07 SP, France
Tel: +33–1–42–19–1675
Fax: +33–1–42–19–1836

Jacques Theys
Chef du Bureau de la Coordination interministérielle
Direction générale de l'Administration et du Développement
Ministère de l'Environnement
20 Avenue de Ségur, 75302 Paris 07 SP, France
Tel: +33–1–40–816323
Fax: +22–1–40–816396

French Sustainable Development Commission
Michel Hors
Secrétaire de la Commission française du Développement durable
Commissariat général du Plan
5 Rue Casimir Périer, 75007 Paris, France
Tel: +33–1–45–565112

Latvia

National Environmental Policy Plan
Valts Vilnitis
Director, Environmental Protection Department
Ministry of Environmental Protection and Regional Development
25 Peldu str, LV–1494 Riga, Latvia
Tel: +371–2–371–2–212696
Fax: +371–2–371–7–820442

The Netherlands

National Environmental Policy Plan
Gerard Keijzers (Director, Strategic Planning)
Directorate for Strategic Planning
Ministry of Housing, Spatial Planning and the Environment
PO Box 30945, Code 675
2500 GX, The Hague, The Netherlands
Tel: +31–70–3394006
Fax: +31–70–3391233

Action Plan: Sustainable Netherlands (and Sustainable Europe Project)
Maria Buitenkamp
Milieudefensie (Friends of the Earth Netherlands)
Damrak 26, PO Box 19199, 1000 GD Amsterdam
The Hague, The Netherlands
Tel: +31–20–6221366
Fax: +31 20–6275602

New Zealand

Resource Management Act, and Environment 2010 Strategy
Lindsay Gow
Deputy Secretary for the Environment
Ministry for the Environment
PO Box 10362, Wellington, New Zealand
Tel: +64–4–473–4090
Fax: +64–4–471–0195

Norway

Reports to Parliament follow-up to WCED, and on UNCED
Paul Hofseth, Special Adviser
Ministry of the Environment
PO Box 8013, 0030 Oslo, Norway
Tel: +47–22–34–9090
Fax: +47–22–34–9560

Poland

National Environmental Policy
Marek Haliniak
Director, Department of Environmental Policy
Ministry of Environmental Protection, Natural Resources and Forestry, Warsaw

Tomasz Zylicz
Warsaw Ecological Economics Centre
Economics Department, Warsaw University
ul Dluga 44/50, 00–241 Warsaw, Poland
Fax: +48–22–254784

Sweden

Government Bill 1993/4:111: Towards Sustainable Development in Sweden
Sture Persson
Assistant Under Secretary
Ministry of the Environment
S–103 33, Stockholm, Sweden
Tel: +46–8–405–1000
Fax: +46–8–21–1690

Enviro '93: Action Programme of the Swedish Environmental Protection Agency
Mats Olson
Deputy Director-General
Swedish Environmental Protection Agency
S–171 85, Solna, Sweden
Tel: +46–8–799–1000
Fax: +46–8–29–2382

United Kingdom

Sustainable Development: The UK Strategy
John Stevens
Head of Environmental Protection, Central Division
Department of the Environment
Romney House

43 Marsham Street, London SW1P 3PY, UK
Tel: +44–171–276–8599
Fax: +44–171–276–6430

United States of America

President's Council on Sustainable Development

Molly Harriss Olson
Executive Director
President's Council on Sustainable Development
730 Jackson Place NW, Washington DC
Tel: +1–202–408–5296
Fax: +1–202–408–6839

[Note: The PCSD offices were closed at the end of March 1996]
Alternatively contact:

Jonathan Lash
(Co-chairman, President's Council on Sustainable Development)
President, World Resources Institute
1709 New York Avenue NW, Suite 700
Washington DC
Tel: +1–202–638–6300
Fax: +1–202–638–0036

National Environmental Goals Project

Frederick W (Derry) Allen
Director, Office of Strategic Planning and Environmental Data
Office of Policy, Planning and Evaluation
US Environmental Protection Agency
401 M Street SW (2161), Washington DC 20460
Tel: +1–202–260–4028
Fax: +1–202–260–0275

Acronyms

ACOA	Atlantic Canada Opportunities Agency
ADEME	Agence de l'Environnement et de la maîtrise de l'Énergie
AEPS	Arctic Environmental Protection Strategy
AIDAB	Australian International Development Assistance Bureau
ALGA	Australian Local Government Association
APSN	Action Plan: Sustainable Netherlands
BAP	Biodiversity action plan
BC	British Columbia
BCRTEE	British Columbia Round Table on the Environment and the Economy
BEP	Basic Environment Plan
CCME	Canadian Council of Ministers of the Environment
CEC	Commission of the European Communities
CEE	Cabinet Committee on the Environment
CEE	Central and Eastern Europe
CLGF	Commonwealth Local Government Forum
CoA	Commonwealth of Australia
CORE	Commission on Resources and the Environment
CSD	Commission for Sustainable Development
CTED	Department of Community, Trade and Economic Development
DoE	Department of the Environment
DEST	Department of Environment, Sport and Territories (Australia)
DG	Directorate-General
DIREN	Direction régionale de l'Environnement
EAJ	Environment Agency of Japan
EAP	Environmental Action Programme
EC	European Commission
ECIN	Education and Community Involvement Network
EEA	European Environment Agency
EEC	European Economic Community
EFTA	European Free Trade Area
EIA	Environmental impact assessment
EMAS	Eco-Management and Audit Scheme
ENGO	Ecological non-governmental organization

EPA	Environmental Protection Agency
EPD	Environmental Protection Department
EPE	Environmental Programme for Europe
ERP	Environmental Resource Program
ESD	Ecologically Sustainable Development
ESDSC	ESD Steering Committee
EU	European Union
FCSD	French Commission for Sustainable Development
G7	the group of seven 'leading' industrial nations
GDP	Gross domestic product
GEF	Global Environmental Facility
GIS	Geographical information system
GLOBE	Go Local for a Better Environment
GLWQA	Great Lakes Water Quality Agreements
GNI	Gross national income
GNP	Gross national product
GP	General practitioner
ICESD	Intergovernmental Committee for Ecologically Sustainable Development
ICLEI	International Council of Local Environmental Initiatives
IDRC	International Development Research Centre
IFEN	Institut français de l'Environnement
IGAE	Intergovernmental Agreement on the Environment
IIED	International Institute for Environment and Development
IISD	International Institute for Sustainable Development
IJC	International Joint Commission
IMPEL	implementation and enforcement of legislation
IMSAR	Institute for Environment and Systems Analysis
INGP	International Network of Green Planners
ISD	Institute for Sustainable Development (Poland)
LA	Local Authority
Ind	Industry
ISO	International Standards Organization
IUCN	International Union for the Conservation of Nature
LA21	Local Agenda 21
LGIB	Local Government International Bureau
LGMB	Local Government Management Board
LRTAP	Long Range Transboundary Air Pollution
MARN	Ministry of the Environment and Natural Resources (Portugal)
MEPRD	Ministry of Environmental Protection and Regional Development (Latvia)
MFA	Ministry of Foreign Affairs
MfE	Ministry for the Environment (New Zealand)
MoE	Ministry of the Environment
MS	Member States
NAFTA	North American Free Trade Agreement
NCS	National Conservation Strategy
NEAP	National Environmental Action Programme

NFEP	National Foundation for Environmental Protection (Poland)
NEP	National Environmental Policy (Poland)
NEPA	National Environmental Protection Agency
NEPC	National Environmental Protection Council
NEPP	National Environmental Policy Plan (Netherlands)
NEPPL	National Environmental Policy Plan for Latvia
NERI	National Environmental Research Institute
NESDA	Network for Environmental Sustainable Development in Africa
NGO	non-governmental organization
NRTEE	National Round Table on the Environment and the Economy (Canada)
NSDS	National Sustainable Development Strategy
NSESD	National Strategy for Ecologically Sustainable Development (Australia)
NUP	Austrian National Environmental Plan (National Umwelt Plan)
NZMfE	New Zealand Ministry for the Environment
ODA	Overseas Development Administration (UK)
OECD	Organization for Economic Cooperation and Development
OPPE	Office of Policy, Planning and Evaluation
PAH	Polycyclic aromatic hydrocarbons
PCB	Poly chlorinated biphenol
PCEQ	President's Commission on Environmental Quality
PCSD	President's Council on Sustainable Development (USA)
PEGE	Partnership for Economic Growth and the Environment
PNE	Plan national pour l'Environnement (France)
RAC	Resource Assessment Commission
RAP	Remedial Action Plan
REC	Regional Environment Centre for Central and Eastern Europe
RIVM	National Institute of Public Health and Environmental Protection (the Netherlands)
RMA	Resource Management Act (New Zealand)
SEPP	State Environmental Planning Policies
SLMS	Sustainable Land Management Strategy
SME	Small and medium-sized enterprise
SO_2	Sulphur dioxide
SWMS	Sustainable Water Management Strategy
T&D	Towns and Development
TFAP	Tropical Forestry Action Plan
ToR	Terms of reference
UNA	United Nations Association
UNCED	United Nations Conference on Environment and Development
UNCSD	United Nations Commission on Sustainable Development
UNDP	United Nations Development Programme
UNECE	United Nations Economic Commission for Europe
UNEP	United Nations Environment Programme
VAT	Value added tax
VREN	Vale Royal Environment Network
VROM	Ministry of Housing, Spatial Planning and the Environment (the Netherlands)

WCED	World Commission on Environment and Development
WHO	World Health Organization
WMI	Whitehorse Mining Initiative (Canada)
WWF	World Wide Fund for Nature

References

AIDAB (1991) *Ecologically Sustainable Development in International Development Cooperation: An Interim Policy Statement*, Australian International Development Assistance Bureau, Canberra

ANZECC (undated) *Draft National Strategy for the Conservation of Australia's Biological Diversity*, prepared by the Australian and New Zealand Environment and Conservation Council Task Force on Biological Diversity

ARMCANZ/ANZECC (in preparation) *National Water Quality Management Strategy*, Agriculture and Resource Management Council of Australia and New Zealand, and the Australian and New Zealand Environment and Conservation Council

AustMoE (1995) *Austria: National Environmental Plan*, Federal Ministry of the Environment, Vienna

Bass, S M J and D B Dalal-Clayton (1995) *Small Island States and Sustainable Development: Strategic Issues and Experience*, Environmental Planning Issues No 8, International Institute for Environment and Development, London

Bass, S M J, D B Dalal-Clayton and J Pretty (1995) *Participation in Strategies for Sustainable Development*, Environmental Planning Issues No 7, International Institute for Environment and Development, London

BCRTEE (1992) *Towards a Strategy for Sustainability*, British Columbia Round Table on the Environment and the Economy, Victoria

Bear, D (1994) *Public Responses to the President's Council on Sustainable Development Draft Vision Statement and Principles for Sustainable Development, Report to the Task Force on Principles, Goals and Definition*, Labat-Anderson Incorporated, USA, 27 October

Biodiversity Working Group (1994) *Canadian Biodiversity Strategy: Canada's Response to the Convention on Biological Diversity*, Report of the Federal–Provincial–Territorial Biodiversity Working Group, November, Ottawa

Campbell, A (1992) *Landcare in Australia*, Third Annual Report, National Landcare Facilitator, National Soil Conservation Program, Canberra

— (1994) *Community First: Landcare in Australia*, Gatekeeper Series No 42, Sustainable Agriculture Programme, International Institute for Environment and Development, London

Canada/Mexico/USA (1993) *North American Agreement on Environmental Cooperation*, between the governments of Canada, Mexico and the USA, final draft, 13 September

Canadian Liberal Party (1993) *Creating Opportunities: The Liberal Plan for Canada*, Liberal Party, Ottawa

Carew-Reid, J, R Prescott-Allen, S Bass and D B Dalal-Clayton (1994) *Strategies for National Sustainable Development: A Handbook for their Planning and Implementation*, IIED/IUCN in association with Earthscan Publications Ltd, London and Gland

CEC (1992a) *Towards Sustainability: A European Community Programme of Policy and Action in Relation to the Environment and Sustainable Development*, Commission of the European Communities, Cm (92) 23/Final/II, Brussels

— (1992b) *The State of the Environment in the European Community: Overview*, accompanying document to the proposal from the CEC for a community programme of policy and action in relation to the environment and sustainable development, Commission of the European Communities, Brussels

— (1993a) 'Growth, Competitiveness, Employment: The Challenges and Ways Forward into the 21st Century', *Bulletin of the European Communities*, White Paper, supplement 6/93, Commission of the European Communities, Brussels

— (1993b) 'The Future Development of the Common Transport Policy: A Global Approach to the Construction of a Community Framework for Sustainable Mobility', *Bulletin of the European Communities*, Supplement 3/93, Commission of the European Communities, Brussels

— (1994a) *Interim Report of Implementation Towards Sustainability*, Commission of the European Communities, Brussels, November

— (1994b) *Fourth Framework Programme (1994–1998) Amounts and Breakdowns*, CORDIS focus, RTD–Results, Commission of the European Communities, Brussels, 20 May

— (1995) *An Energy Policy for the European Union*, White Paper, COM(95)682, Commission of the European Communities, Brussels

— (1996a) *Progress Report from the Commission on the Implementation of the European Community Programme of Policy and Action in Relation to the Environment and Sustainable Development 'Towards Sustainability'*, COM(95) 624 final, Commission of the European Communities, Brussels

— (1996b) 'Proposal for a European Parliament and Council Decision on the Review of the European Community Programme of Policy and Action in Relation to the Environment and Sustainable Development "Towards Sustainability"', COM (95) 647 final, 96/0027 (COD), Commission of the European Communities, Brussels, January

Christensen N, H Paaby and J Holten-Andersen (eds) (1994) *Environment and Society: A Review of Environmental Development in Denmark*, Technical Report No 108, National Environmental Research Institute, Roskilde, Denmark, May

Clement, P (1993) *National Overview of Strategic Development in Canada's Provinces and Territories*, Working Paper No 18, National Round Table on the Environment and the Economy, Ottawa

CoA (1991) *Ecologically Sustainable Development Working Groups, Final Report – Executive Summons*, Commonwealth of Australia, Australian Government Publishing Service, Canberra

— (1992a) *National Strategy for Ecologically Sustainable Development*, Commonwealth of Australia, Australian Government Publishing Service, Canberra

— (1992b) *National Greenhouse Response Strategy*, Commonwealth of Australia, Australian Government Publishing Service, Canberra, December

— (1992c) *National Forestry Statement: A New Focus for Australia's Forests*, Commonwealth of Australian Government Publishing Services, Canberra

Commissariat général du Plan (1995) *Synthesis: The European Conference of the National Commissions on Sustainable Development*, 19–21 January, Courchevel, France,

Commission française du Développement durable, Commissariat général du Plan, Paris

Dalal-Clayton, D B, S Bass, B Sadler, K Thomson, R Sandbrook, N Robins and R Hughes (1994) *National Sustainable Development Strategies: Experience and Dilemmas*, Environmental Planning Issues No 6, IIED, London

DanMoE (1991) *The State of the Environment in Denmark*, Ministry of the Environment, Copenhagen, Denmark

— (1994) *The Danish Environmental Strategy*, Ministry of the Environment, Copenhagen, Denmark, May

— (1995a) *Local Agenda 21: An Introduction to Counties and Municipalities*, 2 March, Danish Ministry of the Environment in collaboration with the Association of Municipalities and the Association of Counties, 2 March

— (1995b) *Denmark's Nature and Environment Policy 1995: Summary Report*, Danish Ministry of the Environment, Copenhagen, June

DanSIS (1988) *Environment and Development: The Danish Government's Action Plan*, follow up to the recommendations in the report of the World Commission on Environment and Development and the UN Environmental Perspective to the year 2000, Danish State Information Service, Copenhagen, Denmark, December

DEST (1994) *Australia's National Report 1994 for the United Nations Commission on Sustainable Development*, Department of Environment, Sport and Territories, Canberra

Doering, R L (1993) *Canadian Round Tables on the Environment and the Economy: Their History, Form and Function*, Working Paper No 14, March (revised September), National Round Table on the Environment and the Economy, Ottawa

EAJ (1994) *The Basic Environment Plan*, Environment Agency of Japan, Tokyo, December

— (1995) *The Basic Environment Plan: An Outline*, Environment Agency of Japan, Tokyo (April 1995),

Earth Council (1995) *Directory of National Councils for Sustainable Development – or Similar Initiatives* (2nd edition, April 1995), The Earth Council, San José, Costa Rica

EEA (1994) *Europe's Environment. The Dobris Assessment: An Overview*, European Environment Agency, Copenhagen, October

EEB (1995) *Review of the Fifth Environmental Action Programme*, European Environmental Bureau, Brussels

Environment Canada (1990) *A Framework for Discussion on the Environment*, Catalogue No En21–86/1990, Environment Canada, Hull, Quebec

ERM (1994a) *Approaches to Strategic Environmental Management*, Paper I prepared for the Ministry of Housing, Spatial Planning and Environment in the Netherlands by Environmental Resources Management (UK), and presented to the first meeting of the International Network of Green Planners, 30 March–1 April 1994, Maastricht, the Netherlands

— (1994b) *Developing Plans and Strategies*, Paper II prepared for the Ministry of Housing, Spatial Planning and Environment in the Netherlands by Environmental Resources Management (UK), and presented to the first meeting of the International Network of Green Planners, 30 March–1 April 1994, Maastricht, the Netherlands

— (1994c) *Implementing Plans and Strategies*, Paper III prepared for the Ministry of Housing, Spatial Planning and Environment in the Netherlands by Environmental Resources Management (UK), and presented to the first meeting of the

International Network of Green Planners, 30 March–1 April 1994, Maastricht, the Netherlands

— (1994d) *Moving from Environmental to Sustainable Development Strategies*, Paper IV prepared for the Ministry of Housing, Spatial Planning and Environment in the Netherlands by Environmental Resources Management (UK), and presented to the first meeting of the International Network of Green Planners, 30 March–1 April 1994, Maastricht, the Netherlands

— (1995) *Implementing Green Plans: Ideas and Experience from Surface Water Management*, Paper VI prepared for the Ministry of Housing, Spatial Planning and Environment in the Netherlands by Environmental Resources Management (UK), and presented to the second meeting of the International Network of Green Planners, 1–3 June, 1995, San Francisco

Esambert, B (1995) *The Conditions and the Means for Sustainable Development* (English), Commission française du Développement durable, Commissariat général du Plan, Paris

ESDSC (1993) *Summary Report of the Implementation of the National Strategy for Ecologically Sustainable Development*, prepared by the ESD Steering Committee, Department of the Prime Minister and Cabinet, Canberra

FCSD (1995) *The European Conference of the National Commissions on Sustainable Development*, Courchevel, 19–20 January, French Commission on Sustainable Development, Commissariat général du Plan, Paris

FoE Europe (1995) *Recommendations from the Sustainable Europe Conference*, 17 November, Friends of the Earth Europe, Brussels

French MoE (1990a) *Plan national pour l'Environnement: Rapport préliminaire en vue du Débat d'Orientation*, L Chabson and J Theys (eds), Secrétariat d'État auprès du Premier Ministre (SEPM), Chargé de l'Environnement et de la Prévention des Risques technologiques et naturels majeurs, Ministry of the Environment, Paris, June

— (1990b) *Plan national pour l'Environnement*, Supplément à Environnement Actualité No 122, Secrétariat d'État auprès du Premier Ministre (SEPM), Chargé de l'Environnement et de la Prévention des Risques technologiques et naturels majeurs (Ministry of the Environment), Neilly-sur-Seine, September

— (1990c) *The National Plan for the Environment: A Summary (English)*, Ministry of the Environment, Paris, November

— (1994) *Chartes pour l'Environnement, Livret de Présentation*, Direction de la Nature et des Paysages, Ministère de l'Environnement, Paris

— (1995) *Extract of the Report of France to the United Nations Commission for Sustainable Development* (in English), Ministry of the Environment, Paris

GCCEPU (1993) *Sustainable Development Seminar*, 18–20 March, Green College Centre for Environmental Policy and Understanding, Oxford

GMoE (1994) *Environment 1994: German Strategy for Sustainable Development – Summary*, German Federal Ministry for the Environment, Bonn

Government of Canada (1990a) *Canada's Green Plan*, Catalogue No En21–94/1990E, Minister of Supply and Services, Canada

— (1990b) *A Report on the Green Plan Consultations*, August, Catalogue No En21–90/1990E, Minister of Supply and Services, Canada

— (1995) *A Guide to Green Government*, Catalogue No En21–136/1995E, Ministry of Supply and Services, Canada

Government of Japan (1993) *National Action Plan for Agenda 21*, The Government of Japan, December

Government of New Zealand (undated) *Path to 2010: Securing a Future for New Zealanders to Share.* Enquiries to Department of the Prime Minister and Cabinet, Wellington, New Zealand

Grundy, K J (1993) *Sustainable Development: A New Zealand Perspective*, Publication No 3, Environmental Policy and Management Centre, University of Otago, Dunedin, New Zealand

Gummer, John (1994) Speech to ERM Forum at the Royal Society of Arts, 20 December, London

Hill, J (1992) 'National Sustainability Strategies: A Comparative Review of the Status of Five Countries – Canada, France, the Netherlands, Norway and UK', paper presented to the Department of the Environment Seminar on Sustainable Development, Oxford, 18–20 March, The Green Alliance, London

HMSO (1990) *This Common Inheritance: Britain's Environmental Strategy*, Cm 1200, Her Majesty's Stationery Office, London, September

— (1991) *This Common Inheritance: The First Year Report*, Cm 1655, Her Majesty's Stationery Office, London, September

— (1992) *This Common Inheritance: The Second Year Report*, Cm 2068, Her Majesty's Stationery Office, London, October

— (1994a) *Sustainable Development: The UK Strategy*, Cm 2426, Her Majesty's Stationery Office, London

— (1994b) *Climate Change: The UK Programme*, Cm 2427, Her Majesty's Stationery Office, London

— (1994c) *Biodiversity: The UK Action Plan*, Cm 2428, Her Majesty's Stationery Office, London

— (1994d) *Sustainable Forestry: The UK Programme*, Cm 2429, Her Majesty's Stationery Office, London

— (1994e) *This Common Inheritance: The Third Year Report*, Cm 2549, Her Majesty's Stationery Office, London, May

— (1995) *This Common Inheritance: UK Annual Report*, Cm 2822, Her Majesty's Stationery Office, London, March

Hofseth, P (1993) 'Environment Plans', Paper presented to Environment Canada Meeting on Green Plans

IDoE (1990) *Environmental Action Programme*, Irish Department of the Environment, Dublin

— (1995a) *Moving Towards Sustainability: A Review of Recent Environmental Policy and Papers*, Irish Department of the Environment, Dublin, May

— (1995b) *Sustainable Development and Local Authorities: Guidelines on Local Agenda 21*, Irish Department of the Environment, Dublin, June

IFEN (1995a) *L'État de l'Environnement dans les Territoires françaises du Pacific Sud: La Nouvelle Calédonie*, Institut français de l'Environnement, Orléans, March

— (1995b) *L'État de l'Environnement dans les Territoires françaises du Pacific Sud: Wallis et Futuna*, Institut français de l'Environnement, Orléans, March

— (1995c) *L'État de l'Environnement dans les Territoires françaises du Pacific Sud: Polynesie française (Tahiti)*, Institut français de l'Environnement, Orléans, March

IPAMB (1995) *The National Environmental Plan: Abridgement*, Ministério do Ambiente e Recursos Naturais, Instituto de Promoçâo Ambiental, Lisbon, June

ISD (1993) *Green Lungs of Europe – Belarus, Estonia, Latvia, Lithuania, Poland, Russia and Ukraine – Outline Concept*, Institute of Sustainable Development, Warsaw, Poland, February

IUCN (1993a) *Regional Review of Strategies for Sustainability in South and East Asia, Volume 1*, review draft, World Conservation Union (IUCN), Gland, Switzerland

— (1993b) *Regional Review of Strategies for Sustainability in Africa, Volume 1*, review draft, World Conservation Union (IUCN), Gland, Switzerland

— (1993c) *The IUCN Joint Programme on Strategies for Sustainability: Activities and Lessons Learned*, World Conservation Union (IUCN), Gland, Switzerland

IUCN/UNEP/WWF (1980) *World Conservation Strategy: Living Resource Conservation for Sustainable Development*, International Union for Conservation of Nature (IUCN), United Nations Environmental Programme (UNEP) and World Wide Fund for Nature (WWF), Gland, Switzerland

Lawrence, G (1994) 'Sustainable Seattle, USA', in UKDoE, *Partnerships in Practice*, Department of the Environment, UK

LNV/VROM (1995) *Structure Plan for the Rural Areas in the Netherlands*, Ministry of Agriculture, Nature Management and Fisheries (LNV) and Ministry of Housing, Spatial Planning and the Environment (VROM), The Hague

MEPRD (1995) *National Environmental Policy Plan for Latvia* (English translation), Ministry of Environmental Protection and Regional Development, Riga, Latvia

Milieudefensie (1992) *Action Plan: Sustainable Netherlands*, Milieudefensie, Amsterdam (English translation, 1993)

NFEP (1991) *Outline of a Regional Policy for the Eco-Region 'Green Lungs of Poland'*, prepared by the 'Green Lungs of Poland' Division for Ecological Services, National Foundation for Environmental Protection, Suwalki, Poland

Nguyen, Y-Lang and T Roberts (1994) *State Progress Toward Sustainable Development*, draft, November, Center for Policy Alternatives, Washington DC

NorMoE (1989) *Report to the Storting No 46 (1988–9), Environment and Development: Programme for Norway's Follow-Up of the Report of the World Commission on Environment and Development*, Norwegian Ministry of Environment, Oslo

— (1992a) *Report to the Storting No 64 (1991–92), Concerning Norway's Implementation of the North Sea Declarations*, Norwegian Ministry of Environment, Oslo

— (1992b) *Report to the Storting No 13 (1992–3) on the UN Conference on Environment and Development, Rio de Janeiro*, Norwegian Ministry of Environment, Oslo

— (1994a) *Report on the Symposium on Sustainable Consumption*, 19–20 January, Oslo, Norway, Norwegian Ministry of Environment, Oslo

— (1994b) *Report on Norway's National Communication under the Framework Convention on Climate Change*, Norwegian Ministry of Environment, Oslo

— (1995) *Oslo Ministerial Roundtable: Conference on Sustainable Production and Consumption*, Norwegian Ministry of Environment, Oslo

NRTEE/ParticipACTION (1994) *SustainABILITY: A National Communications Program in Support of Sustainable Development*, presented to the Government of Canada, December, National Round Table on the Environment and the Economy, Ottawa, and ParticipACTION, Toronto

NZMfE (1993a) *Cleaner Production at Work: Case Studies from New Zealand*, Ministry for the Environment, Wellington, New Zealand, July

— (1993b) *Business and the Environment: A Developing Partnership. The Report of a National Conference of Business Leaders*, Ministry for the Environment, Wellington, New Zealand, September

— (1994a) *Environment 2010 Strategy: A Statement of the Government's Strategy on the Environment*, Ministry for the Environment, Wellington, New Zealand, October

— (1994b) *Summary of Draft Environment 2010 Strategy*, Ministry for the Environment, Wellington, New Zealand, October

— (1995) *Environment 2010 Strategy*, Ministry for the Environment, Wellington, New Zealand

OECD (1995a) *Planning for Sustainable Development: Country Experiences*, OECD, Paris

— (1995b), *OECD Environmental Performance Review of the Netherlands*, OECD, Paris
— (1995c) *Environmental Performance Review of Poland*, Organization for Economic Cooperation and Development, Paris
PCSD (1996) *Sustainable America: A New Consensus for Prosperity, Opportunity, and a Healthy Environment for the Future*, President's Council on Sustainable Development, Washington DC, February
Plowman, J (1993) Presentation to the European Environmental Bureau Round Table on the Implementation of the EC Fifth Environmental Action Programme on the Environment, 2 March, London
PolMEP (1990) *National Environmental Policy* (adopted by Parliament in 1991), Polish Ministry of Environmental Protection, Natural Resources and Forestry, Warsaw
— (1994a) *NEP Implementation Plan to Year 2000: A Summary* (presented to the Council of Ministers, September 1994), prepared by MOSZNIL, POIS, IOS, IMUZ from Warsaw Polytechnic and GEA, EKO–EFEKT under coordination of J Zurek, Polish Ministry of Environmental Protection, Natural Resources and Tourism, Warsaw
— (1994b) *Statement of Progress of NEP Implementation 1991–1993* (presented to the Council of Ministers, September), prepared by MOSZNIL, POIS, IOS, Warsaw Polytechnic under coordination of J Zurek, Polish Ministry of Environment, Natural Resources and Tourism, Warsaw
Projet de société (1993) *Prospectus for the Projet de Société*, National Round Table on the Environment and the Economy, Ottawa, Canada, updated 1994
— (May, August and December 1994, January and April 1995) *Newsletters*, National Round Table on the Environment and the Economy, Ottawa
— (1994a) *Planning for a Sustainable Future: Projet de Société – A Partnership for Change. Prospectus*, National Round Table on the Environment and the Economy, Ottawa, May
— (1994b) 'Choices for Transition to Sustainability: Vol 5, Towards a National Sustainable Development Strategy for Canada', draft, November, National Round Table on the Environment and the Economy, Ottawa, Canada
— (1995a) *Newsletter*, 1 (4), April, National Round Table on the Environment and the Economy, Ottawa, Canada
— (1995b) *Canadian Choices for Transitions to Sustainability*, vol 5, Towards a National Sustainable Development Strategy for Canada, final draft, May, National Round Table on the Environment and the Economy, Ottawa
REC (1994) *Summary of the Environmental Action Programme for Central and Eastern Europe*, document endorsed by the Ministerial Conference in Lucerne, Switzerland, 28–30 April, 1993, Regional Environmental Centre for Central and Eastern Europe, Budapest (March 1994)
— (1995) *Status of National Environmental Action Programs in Central and Eastern Europe*, Regional Environmental Centre for Central and Eastern Europe, Budapest
Redclift, M (1992) 'The Meaning of Sustainable Development', in IIED, *Sustainable Development in Peru: A Review of a Document from the Pachamama Society*, International Institute for Environment and Development, London
RIVM (1989) *Concern for Tomorrow*, National Institute of Public Health and Environmental Protection (RIVM), Bilthoven
— (1991) *Zorgen voor Morgen: Nationale Milieuverkenning 1985–2101 (Caring for Tomorrow, National Environmental Outlook 1985–2010)*, National Institute of Public Health and Environmental Protection (RIVM), Alphen aan de Rijn, Samson HD Tjeenk Willink
Rowley, J (1993) 'Confounding the Cynics', *People and Planet*, 2 (4), p3

SAB (1990) *Reducing Risk: Setting Priorities and Strategies for Environmental Protection*, Science Advisory Board, Washington DC, September

Sadler, B (1996) 'Sustainability Strategies and Green Planning: Recent Canadian and International Experience', in A Dale and J Robinson (eds), *Achieving Sustainable Development*, University of British Columbia Press, Vancouver

Scanlon, J E L (1995) 'The Need for a Legislative Framework to Transform the Concept of Sustainable Development into a Reality: New Zealand Legislates: Australia Procrastinates', Master of Laws (Environmental) dissertation, Australian Centre for Environmental Law, Adelaide, April

Silveira, D (1995) *Facilitating People's Initiatives Programme: Summary Report and Recommendations*, report to Earth Council Institute meeting, 18–19 September, Earth Council, San José, Costa Rica

Stanners, D and P Bourdeau (eds) (1995) *Europe's Environment: The Dobris Assessment*, European Environment Agency, Copenhagen, Earthscan Publications Ltd, London

Stevens, J (1995) *The UK's Strategy for Sustainable Development*, a presentation by the UK Government to the UN Commission on Sustainable Development, April, New York

Sustainable Seattle (1993) *1993 Indicators of Sustainable Community*, Sustainable Seattle

SwedEPA (1993) *An Environmentally Adapted Society: The Action Programme of the Environmental Protection Agency: Environment* 93, Swedish Environmental Protection Agency, Stockholm, June

— (1994) *Strategy for Sustainable Development: Summary of Proposals for a Swedish Programme, Enviro '93*, Swedish Environment Protection Agency, Solna, Sweden

SwedMoE (1994) *Towards Sustainable Development in Sweden: Implementation of the Resolutions of the United Nations Conference on Environment and Development*, Swedish Government Bill 1993/94:111 (presented to Parliament on 9 December 1993, adopted late April 1994), Swedish Ministry of the Environment and Natural Resources, Stockholm

— (1995) *The Environment: Our Common Responsibility. (1) Environmental Policy: Focus and Strategy; (2) Report on the State of the Environment in Sweden*, government communication 1994/5:120, presented to Parliament on 20 December 1994, Swedish Ministry of the Environment, Stockholm

Task Force on Economic Instruments and Disincentives to Sound Environmental Practices (1994) *Economic Instruments and Disincentives to Sound Environmental Practices. Final Report of the Task Force*, Environment Canada, Hull, Quebec, November

Toner, G (1994) 'The Green Plan: From Great Expectations to Eco-Backtracking . . . to Revitalization?' in S D Phillips (ed) (1994) *How Ottawa Spends 1994–95*, Carleton University Press, Ottawa

UKDoE (1993) *UK Strategy for Sustainable Development: Consultation Paper*, Department of the Environment, London, July

UNA–UK (1995) *Towards Local Sustainability: A Review of Current Activity on Local Agenda 21 in the UK*, United Nations Association–UK and Community Development Foundation, London

UNCED (1992) *Agenda 21*, UNCED, United Nations General Assembly, New York

UNECE (1995) *Environmental Programme for Europe*, submitted to the Ministerial Conference 'Environment for Europe', 23–5 October 1995, Sofia, Bulgaria, by the ECE Working Group of Senior Governmental Officials, United Nations Economic Commission for Europe

Upton, S D (1995) *Purpose and Principle in The Resource Management Act*, The Stace Hammond Grace Lecture, 26 May 1995, University of Waikato, New Zealand

USEPA (1994) 'EPA's National Environmental Goals Project', project description notes, November, United States Environmental Protection Agency, Washington DC, mimeo

— (1995) *Proposed Environmental Goals for America with Benchmarks for the Year 2005: Summary*, Doc No EPA–230–D–95–001, February, United States Environmental Protection Agency, Washington DC

VROM (1989) *National Environmental Policy Plan*, Department for Information and International Relations, Ministry of Housing, Spatial Planning and the Environment, The Hague

— (1991) *Netherlands National Report to UNCED 1992*, Doc No VROM 91336/a/8–91, Department for Information and International Relations, Ministry of Housing, Spatial Planning and the Environment, The Hague, July

— (1993a) *The Netherlands' National Environment Policy Plan 2*, Doc No VROM 93561/b/4–94, Department for Information and International Relations, Ministry of Housing, Spatial Planning and the Environment (VROM), The Hague

— (1993b) *National Planning for Sustainable Development: The Netherlands Experience*, Doc No VROM 93523/h/10–93, Directorate for Strategic Planning, Ministry of Housing, Spatial Planning and the Environment, The Hague

— (1994a) *Towards a Sustainable Netherlands: Environment Policy Development and Implementation*, Doc No VROM 93573/b/3–94, Department for Information and International Relations, Ministry of Housing, Spatial Planning and the Environment, The Hague, January

— (1994b) *Environmental Policy of the Netherlands*, Doc No VROM 93613/B/12–93, Department for Information and International Relations, Ministry of Housing, Spatial Planning and the Environment (VROM), The Hague, March

WCED (1987) *Our Common Future*, Report of the World Commission on Environment and Development, Oxford University Press, Oxford, England

Weichel, B (1993) *Common Destination: An Analysis of Canadian Strategies for Sustainable Development*, unpublished report, Canadian Council of Ministers of the Environment, Winnipeg

Weizsäcker, E von (1994) *Earth Politics*, Zed Books, London

World Bank (1994) *Making Development Sustainable*, The World Bank Group and the Environment, Fiscal, The World Bank, Washington DC

— (1995) *National Environmental Strategies: Learning from Experience*, World Bank, Washington DC

Wuppertal Institute (1995) *Towards Sustainable Europe: The Study*, Friends of the Earth Europe, Brussels, February

Zylicz, T (1994) *Poland's National Environmental Policy: Outline of Economic Instruments After 4 Years*, Ecological Economics Centre, Warsaw University, Poland

Further Reading

CFSD (1995) *The European Conference of the National Commissions on Sustainable Development: Synthesis*, Commission française du Developpement durable, Commissariat général du Plan, Paris

CoA (1990) *Ecologically Sustainable Development: A Commonwealth Discussion Paper*, Commonwealth of Australia, Australian Government, Canberra, June

DanMoE (1992) *Environmental Initiatives in the 1990s: Objectives, Principles and Main Strategies*, Danish Ministry of the Environment, Copenhagen

Doern, G B (1991) *Shades of Green: Gauging Canada's Green Plan*, C D Howe Institute, Commentary No 29, Toronto, April

Environment Canada (1994) *Environment Canada Business Plan 1994 to 1999*, Cat.No.En.21–86/1990, Environment Canada, Hull, Quebec

French MoE (1994) *Rapport de la France à la Commission du Développement durable des Nations-Unies (1995)*, Ministry of the Environment, Paris, December

Government of Canada (1992) *Canada's Green Plan and the Earth Summit*, Minister of Supply and Services, Canada

— (1992) *Economic Instruments for Environmental Protection*, Discussion Paper, Cat.No.En.21–119/1992E, Minister of Supply and Services, Canada

— (1994) *Response to the First Report of the House of Commons Standing Committee on Environment and Sustainable Development (Commissioner of the Environment and Sustainable Development*, Office of the Deputy Prime Minister and Minister of the Environment, Hull, Quebec, October

Gow, L (1995) *Implementing Sustainability: New Zealand's Experience with its Resource Management Act*, Address to the World Resource Institute/New Zealand Embassy Seminar, Washington DC, 6 June

Greenberg, P A (1993) *Toward a US Green Plan: Thinking about a US Strategy for Sustainable Development*, Report commissioned by the Resources Renewal Institute, San Francisco

Hofseth, P (1995) 'Norway's Sustainable Development Strategy', in F Kvaloy (ed) *Proceedings of the Workshop on Sustainable Development Strategies: 7–8 June 1994, Oslo*, Department of Development Coordination Programmes, Ministry of Foreign Affairs, Oslo

IFEN (1994) *The Environment in France: 1994–1995 Edition. Key Facts and Figures*, Institut français de l'Environnement, Orléans, March

IOV (1994) *Environment and Development Cooperation: Evaluation of Netherlands Aid Policy with Regard to the Environment, with Special Reference to Burkina Faso,*

Indonesia and Kenya, Summary Evaluation Report, Operations Review Unit (IOV), Ministry of Foreign Affairs, The Hague

Kirton, J (1994) *Sustainable Development as a Focus for Canadian Foreign Policy*, Working Paper No 25, September, National Round Table on the Environment and the Economy, Ottawa

Lampietti, J A and U Subramanian (1995) *Taking Stock of National Environmental Strategies*, Environmental Management Series Paper No 010, Environment Department, The World Bank, Washington DC

LGMB (undated) *Local Agenda 21 Roundtable Guidance Notes*, Local Government Management Board, Luton, UK

Milieudefensie (1994) *Sustainable Netherlands*, Milieudefensie (Friends of the Earth Netherlands), Amsterdam

Netherlands Committee for IUCN (1994) *The Netherlands and the World Ecology*, Netherlands Committee of IUCN, Amsterdam

NorMoE (1993) *Report to the Storting No 31 (1992–93): Regional Planning and Land Use Policy*, Norwegian Ministry of Environment, Oslo

— (1994) *1994 Environmental Policy Statement*, Norwegian Ministry of Environment, Oslo

NZNCC (1987) *New Zealand Conservation Strategy*, New Zealand Nature Conservation Council, Wellington

Projet de société (1993) *Overview of the Projet de société*, vol 1, *Towards a National Sustainable Development Strategy for Canada*, Round Table on the Environment and the Economy, Ottawa, December

— (1993) *Report of the Vision and Process Committee*, vol 2, *Towards a National Sustainable Development Strategy for Canada*, Round Table on the Environment and the Economy, Ottawa, December

— (1993) *Report of the Document and Information Committee*, vol 3, *Towards a National Sustainable Development Strategy for Canada*, Round Table on the Environment and the Economy, Ottawa, December

— (1993) *Progress Report, June–December 1993*, vol 4, *Towards a National Sustainable Development Strategy for Canada*, Round Table on the Environment and the Economy, Ottawa, December

UKDoE (1994) *Sustainable Development: The UK Strategy: Summary Report*, Department of the Environment, London

— (1994) *Sustainable Development: The UK Strategy: An Outline*, Department of the Environment, London

USEPA (1994) *The New Generation of Environmental Protection: A Summary of EPA's Five-Year Strategic Plan*, Doc No EPA 200–2–94–001, United States Environmental Protection Agency, Washington DC, July

VROM (1990) *National Environmental Policy Plan Plus*, Department for Information and International Relations, Ministry of Housing, Spatial Planning and the Environment, The Hague

— (1992) *Environmental Policy in the Netherlands*, Doc No VROM 92533/a/11–92, Department for Information and International Relations, Ministry of Housing, Spatial Planning and the Environment, The Hague

— (1993) *Abridged Version of the Environmental Programme: 1994–1997*, Environmental Strategy Publication Series No 1993/22, Doc No VROM 93672/h/12–93, Department for Information and International Relations, Ministry of Housing, Spatial Planning and the Environment, The Hague

— (1994) *Working with Industry*, Environment Policy in Action No 1, Doc No VROM 94066/b/3–94, Department for Information and International Relations, Ministry of Housing, Spatial Planning and the Environment, The Hague
— (1994) *Working with the Construction Sector*, Environment Policy in Action No 2, Doc No VROM 94067/b/3–94, Department for Information and International Relations, Ministry of Housing, Spatial Planning and the Environment, The Hague
— (1994) *Involving the Consumer*, Environment Policy in Action No 3, Doc No VROM 94068/b/2–94, Department for Information and International Relations, Ministry of Housing, Spatial Planning and the Environment, The Hague
— (1994) *Managing Priority Waste Streams*, Environment Policy in Action No 4, Doc No VROM 94069/b/3–94, Department for Information and International Relations, Ministry of Housing, Spatial Planning and the Environment, The Hague
— (1994) *Managing Environmental Information*, Environment Policy in Action No 5, Doc No VROM 9407/b/3–94, Department for Information and International Relations, Ministry of Housing, Spatial Planning and the Environment, The Hague
— (1994) *Achieving Integration*, Environment Policy in Action No 6, Doc No VROM 9407/b/3–94, Department for Information and International Relations, Ministry of Housing, Spatial Planning and the Environment, The Hague
— (1994) *The Environmental Information Policy of the Netherlands*, Doc No VROM 94117/h/3–94, Department for Information and International Relations, Ministry of Housing, Spatial Planning and the Environment, The Hague
— (1994) *Environmental News from the Netherlands*, No 1, Special edition on the Environmental Policy of the Netherlands, Doc No VROM 93566/b/2–94, Department for Information and International Relations, Ministry of Housing, Spatial Planning and the Environment, The Hague
— (1994) *National Environmental Policy Plan: Summary*, Doc No VROM 04059/b/2–94, Department for Information and International Relations, Ministry of Housing, Spatial Planning and the Environment, The Hague
Wuppertal Institute (1995) *A Viable Germany: A Contribution to Global Sustainable Development*, Wuppertal Institute for Climate, Environment and Energy, Wuppertal

Index

acid rain, 172–3, 183, 190
Action Plan: Sustainable Netherlands (APSN), 13, 18–19, 23, 68, 133, 140, 144–7, 233 249
Advisory Committee on Business and the Environment (UK), 193
Africa, xiii, 201
Agarwal, Anil, 20
Agenda 21, xiii, xv, xix, xx, 3–4, 9, 16, 18–20, 29, 41, 47, 56–8, 65, 67, 72, 86–7, 90, 102, 113–14, 124, 143, 153–5, 166, 172, 174, 182–6, 192, 195–9, 214, 217, 224, 229–30, 232, 238, 240; *see also* Local Agenda 21
Air Quality Agreement (Canada–USA), 97
Albania, 14
Alberta, 55, 61, 90, 106
An Taisce, 239
Arctic, 49
Arctic Environmental Protection Strategy (AEPS), 55, 91, 96; *see also* Circumpolar Conservation Strategy
Association of Community Councils, 196
Association of Counties, 57, 114
Association of Municipalities, 57, 114
Atlantic Canada Opportunities Agency (ACOA), 55, 91
Atlantic Coastal Action Plan, 58, 91
Atlantic Ocean, 46
Auditor General Act (Canada), 98
Australia, xv–xviii, 7, 13, 18, 24, 26–7, 31, 54, 56, 58, 62–6, 75–89, 247; *see also* Council of Australian Governments; National Strategy for Ecologically Sustainable Development; South Australia; Western Australia
Australian Conservation Foundation, 18, 78
Australian International Development Assistance Bureau (AIDAB), 77
Australian Local Government Association (ALGA), 75, 82
Austria, xx, 44, 236–7

Baltic Sea, 180
Basic Environment Law (Japan), 239
Basic Environment Plan (BEP) (Japan), 239–40
Bassetlaw, 198
Belarus, 49, 131, 180
Bergen, 46, 169–71
Berlin, 92, 238
Biodiversity, 92
Biodiversity Action Plan(s), 59, 97, 131, 191, 199
Biodiversity Convention, xviii, 180, 205, 223
biodiversity strategy, 91–2, 164, 166, 205
Blakely, Roger, 149
Board of Agriculture (Sweden), 188, 189; *see also* National Board of Agriculture
Botswana, 4
Bouchard, Lucien, 36–7, 50, 92–4
Boundary Waters Treaty (Canada), 55, 91

Brand NZ (New Zealand), 155
British Columbia, 20, 54–5, 61, 90–1, 106–7
British Gas, 197
Browner, Carol, 220
Brundtland, Gro Harlem, 16, 170
Brundtland Commission/report, 9, 14, 18, 22, 38, 44, 77, 92, 138, 148, 165, 167–71, 191, 208, 223; *see also* World Commission on Environment and Development (WCED)
Brussels, 39, 228, 234, 248
Bulgaria, 14, 46, 48
Bund für Umwelt und Naturschutz Deutchland, 238
Bush, President George, 206, 214, 220
Business Council for Sustainable Development, 214
Buzzelli, David T, 209

Cabinet Ad Hoc Committee on Reform of Local Government and Resource Management Statutes (New Zealand), 148, 158, 160
Cabinet Committee on Enterprise, Industry and Environment (New Zealand), 26
Cabinet Committee on the Environment (CEE) (Canada), 62, 92, 96–7
Cabinet Subcommittee on the Environment (UK), 191–3
Calderdale Metropolitan Borough Council, 200–1
California, 58, 217
  Southern California Council on Environment and Development, 58, 217
Campaign for Our Common Future (*Felleskampanjen*), 171
Canada, xv–xviii, 3–4, 7, 13–14, 18, 20, 24–8, 32–7, 39, 49–50, 54–5, 58, 59–61, 64–6, 75, 90–108, 116, 201–2, 217, 247–8
Canadian Council of Ministers of the Environment (CCME), 33, 100
Canadian Environmental Advisory Council, 106
Canadian Environmental Assessment Act, 49, 98
Canadian Healthy Communities Project, 91
central and eastern Europe, 14, 40–1, 46–8, 63, 70, 72, 126, 131, 174, 180, 238; *see also* eastern Europe
Central Local Government Forum (UK), 193
Centre for Policy Alternatives, 205
Centre for Science and Environment (India), 20
Chabason, Lucien, 117
Chamber of Commerce, 199
Charest, Jean, 100
charter letters, 24, 83
Chattanooga, 205, 212, 216
Chernobyl, 13, 138
Chernomyrdin, Prime Minister Victor, 242
Cheshire, 199
Chicago, 205, 212
Choicework, xvi, 18–19, 101–2, 107
Circumpolar Arctic Environment Protection Strategy, 47, 96; *see also* Finnish Initiative
Circumpolar Conservation Strategy, xviii, 49, 96
Clean Air Act (USA), 97, 218
climate action plan, 110, 205
Climate Change Action Program (Canada), 92
Climate Change Convention, xviii, 92, 240; *see also* UN Framework Convention on Climate Change
climate change plan, 166
Climate Change Task Group, 34
Clinton, President, 13, 206, 208, 214, 220
Commission for Environmental Impact Assessment (Poland), 179
Commission for Sustainable Development (CSD) (in France, Commission française du Développement durable/FCSD), 7, 13, 18, 115, 123–4; (Norway), 170, 172; (Poland), 174; (Sweden), 63, 183; (UN), 4, 87, 124; (USA), 18

Commission on Resources and the Environment (CORE) (Canada), 54
Common Agricultural Policy, 45
Commonwealth Coastal Policy, 89
Commonwealth Local Government Forum (CLGF), 201
communist, 14, 63–4, 175–6, 178–9
Communities Network (UK), 201, 202
conflict resolution, 83, 85–6, 95, 107, 131, 142, 162, 186, 190
Connecticut, 218
Conservation and Development Programme (UK), 39
convention strategies, 59–60
 convention-related strategies, 96, 116, 122, 131, 164, 180, 199
Copenhagen, 46, 248
Council for Europe, 46
Council for Sustainable Development (Poland), 178
Council for Voluntary Service (UK), 199
Council of Australian Governments, 64, 82, 87–8
Council of Ministers (EC), 7, 44, 174, 195, 224, 228, 230–1, 233, 241
Courchevel, 13, 125
covenants, 38, 136, 139–41
Croatia, 14
Croydon, 197
Croydon Environmental Challenge, 197
Czech Republic, 14, 46, 201

Dalal-Clayton, D B, 29, 56
Delphi techniques, 190
Denmark, xv–xviii, 7, 9, 22, 24, 25, 40, 51, 57, 63, 109–14, 248
Department of Commerce (USA), 214
Department of Community, Trade and Economic Development (CTED) (USA), 219
Department of Energy (USA), 205
Department of Environment, Sport and Territories (DEST) (Australia), 79, 82, 88, 247
Department of Primary Industries and Energy (Australia), 82
Department of the Prime Minister and Cabinet (Australia), 82
Department of Renewable Resources (Canada), 55
Department of Trade and Industry (UK), 191
developed countries,68, 70, 71
developing countries, 3–4, 68–72, 196, 211, 241
Detroit, 205
Division of State Cadastres and Natural Resources (Latvia), 128, 129
Dobris Assessment, 46, 49
Doering, Ronald, 34–5
Dow Chemical Company, 206, 209
Dublin, 46, 223, 227
Dutch Antilles, 143

Earth Charter, 217
Earth Day, 179
East Midlands Business and Environment Club (UK), 198
eastern Europe, 6, 22, 47–8, 59–60, 118, 190; *see also* central and eastern Europe
Eco-Management and Audit Scheme (EMAS) (UK), 198
Ecocycle Bill (Sweden), 65, 186
ecological footprints, 5, 8, 20, 97, 107, 122, 164, 172, 183, 199, 228–9, 246
Economic Development Cabinet (USA), 217
Economic Instruments Collaborative (Canada), 34
Education and Community Involvement Network (ECIN) (UK), 197
Ehrlich, Paul, 147
Elf Acquitaine, 115
Energy Agency (Denmark), 40, 113, 128
Energy Association (Latvia), 129
England, xviii; *see also* UK
Enviro '93, xvi, 7, 16, 22–3, 25, 26, 41, 57, 182, 186–90, 251
Environment 2010 Strategy/project/ summit, xviii, 7, 16, 26, 51, 65, 148, 149, 152–7, 160–3, 219, 250

Environment Agency (UK), 195; *see also* European Environment Agency
Environment Canada, 3, 7, 21, 25, 32, 33, 36, 49–51, 90, 92–8, 100, 248
Environment Council (Poland), 177, 225
Environment Forum (UK), 197–8; *see also* Voluntary Sector Environment Forum
Environment Protection Committee (Latvia), 126; *see also* State Committee for Nature Protection
Environmental Action Programme (EAP), xviii, 14, 40, 43, 47–8, 126, 131, 174, 238; *see also* National Environmental Action Programme (NEAP)
  European Community EAP, 223–4
  Fifth EAP, 6, 7, 14, 25–6, 39, 44–5, 58, 66, 70, 195, 223–4, 227–32
  Regional EAP for Central and Eastern Europe, 14, 47–8
environmental charters, 56–7, 122, 199
Environmental Goals Project, 22–3, 26, 41, 49, 51, 205; *see also* National Environmental Goals Project
Environmental Management Act (New Zealand, (1993), 65, 135
Environmental Policy Review Group (European Union), 45, 231–2
Environmental Programme for Europe (EPE), 46, 48
Environmental Protection Act (Canada), 37, 92
Environmental Protection Agency (EPA), 40, in Denmark 110, 112–13, in Ireland, 238, in Sweden, 7, 16, 23, 27, 41, 57, 182–3, 186–8, 251, in the USA, 7, 23, 27, 41, 49, 51, 205, 209, 214, 220, 252; *see also* National Environmental Protection Agency (NEPA)
Environmental Protection Club (Latvia) 129
Environmental Protection Council (Australia), 64; *see also* National Environmental Council (NEPC)
Environmental Protection Department (EPD) (Latvia), 126, 128–9, 249
Environmental Resource Centre (UK), 198
Environmental Resource Program (ERP) (USA), 219
environmental space, 5, 19, 20, 23, 110, 144–7, 228–9, 233–4, 246
environmental targets, 22, 57, 112, 135, 190, 232
Esambert, Bernard, 123–4
Essen, 46
Estonia, 46, 47, 180
European Commission (EC), xi, 7, 39, 44–8, 66, 143, 193–5, 223–33, 248
European Economic Community (EEC), 43–4
European Environment Agency (EEA), 46, 49, 232
European Environmental Bureau, 39, 227–8, 232
European Free Trade Area (EFTA), 46
European Parliament, 44, 223, 227, 229–30, 233
European Union, xi, xiv–xv, xvii, xx, 6, 7, 14, 21, 25–6, 39, 43–6, 70, 110–11, 143, 175, 179, 195, 223–4, 226–30, 232–5, 237, 248; *see also* Treaty of European Union
Eurostat, 46
eutrophication, 21, 128, 134, 137, 187, 190
Everglades, 217
Executive Order No 12852 (USA), 207–8

Farmers' Federation (Australia), 18, 78
Federal Advisory Committee Act (USA), 207–8, 214
Federal Ministry for the Environment, Nature Conservation and Nuclear Safety (Germany), 238
Federal Ministry of Economic Cooperation and Development (Germany), 238
Federation of Nature Parks (France), 38, 117
Federation of Trade Unions (Netherlands), 144
Finance Bill (1994) (Sweden), 65, 186

Finland, 44
Finnish Initiative, *see also* Circumpolar Conservation Strategy, 47, 96
Fisheries Act (New Zealand), 65
Florida, 217
Forestry Act (1991) (Poland), 177
Forum (Norway), 171
Forum on Environment and Development (Germany), 238
Forward Washington, 219
France, xv, 7, 13, 18, 22, 25–6, 38–9, 45, 52, 56, 63, 115–17, 121–5, 228, 248–9
Franklin dam, 14
Fraser Basin Management Program, 55, 91
French Guyana, 124
French Institute of the Environment (IFEN), 124
Friends of the Earth, xvi, 7, 20, 144, 196–7, 229, 233, 249; *see also* Milieudefensie

G7, 195
Gambia, The, 201
Garrett, Paul, 88, 247
Gawlik, Mr Radoslaw, 179
GEN 33 committee, 194
General Consultative Forum (European Union), 45, 229, 231
Germany, xx, 112, 233, 237–8
Global Environmental Facility (GEF), 118, 240
global warming, 21
Gloucestershire, 197
Go Local for a Better Environment (GLOBE), 198
Gore, Vice President, 206, 214, 220
Government Performance and Results Act (USA), 220
Government Training Institute in Kenya, 201
Governor's Commission for a Sustainable South Florida, 217
Great Lakes, 55, 59, 91–2
Great Lakes Water Quality Agreements (GLWQA), 47, 96–7
Greece, 45, 228
green budget, 22, 51, 52, 54, 166, 168, 172
Green College, Oxford, 193
Green Lungs of Europe, 47, 180
Green Lungs of Poland, 47, 180
Green Network of Government Departments, 237
Green Party, 63, 173, 179
Green Plan, Canadian, xvi, xviii, 3, 7, 14, 21, 25–6, 32, 37, 47, 49–51, 54–5, 58, 61–2, 64–5, 90–8, 106, 218, 247
green planning, 3, 5–6, 13–14, 16–17, 21–2, 27, 30, 32, 38, 51, 61, 63–4, 68, 75, 90, 98, 148, 165, 182, 236, 247
greenhouse gas emissions, 76, 92, 110, 157, 180, 192
Guadeloupe, 122
Gulf of Maine (marine) Action Plan, 55, 91
Gummer, John, 43

Hamilton, 56, 59, 90, 91, 164
Hamilton-Wentworth Vision 2000, 59, Vision 2020, 90
Harare, 201
Hawke, Prime Minister Bob, 24, 62–2, 78, 81, 83, 85
Hazardous Substances and Organisms Bill (New Zealand), 65
Health Alliance (UK), 197
Hill, J, 39
Hofseth, Paul, 22, 170, 250
House of Commons Select Committee on the Environment (UK), 202
House of Lords Select Committee on Sustainable Development (UK), 202
Huddersfield University, 200
Hungary, 14, 46

Idaho Sustainability Round-Table Information Forum, 217
Implementation Plan to Year 2000 (Poland), 7, 26, 47, 174, 176, 178–9, 181
India, 20

Institute for Environment and Systems Analysis (IMSAR) (Netherlands), 133
Institute for the Environment (France), 66, 119
Institute of Biology (Latvia), 129
Intergovernmental Agreement on the Environment (IGAE) (Australia), 75–6, 82–3, 86–7
Intergovernmental Committee for Ecologically Sustainable Development (ICESD), 88
integration correspondent(s), 229, 231
International Council of Local Environmental Initiatives (ICLEI), 201–2
International Development Research Centre (IDRC), 33, 100
International Institute for Environment and Development (IIED), xiii, 4–5, 71
International Institute for Sustainable Development (IISD), 33, 100, 105
International Joint Commission (IJC), 55, 91
International Network of Green Planners (INGP), xiii, 3, 4, 72
International Union for Conservation of Nature (IUCN), xiii, xx, 4–5, 46, 67, 71, 180; *see also* World Conservation Union
International Union of Local Authorities, 201
Inuit Circumpolar Conference, 55, 91
Inuit Regional Conservation Strategy, 55, 91
Iowa, 217
Italy, 45, 228

Japan, xx, 239–41
Joint Committee on Sustainable Development (Ireland), 239

Kakadu National Park, 14
Keating, Paul, 62, 85
Kentucky, 217
Kenya, 201; *see also* Government Training Institute in Kenya
Köpenick, 238
Kyrgyz Republic, 48

Labour Organization (Norway), 170
Labour Party (New Zealand), 149
Laender, 237
Lagos, 201
Lalonde, Brice, 25, 63, 115, 117, 121
Landcare (Australia), 18, 77–8; *see also* National Landcare Programme
Lash, Jonathan, 209, 251
Latvia, xv, 7, 14, 22, 25–6, 39–40, 46–7, 59–60, 126–9, 131, 180, 249; *see also* National Environmental Policy Plan for Latvia(NEPPL)
Latvian Fund for Nature, 128–9
Latvian University Ecological Monitoring Centre, 129
Leadership Task Force on Sustainable Economic Development (USA), 219
Leicester, 199
Leiden, 58, 143
Léon City Council, 200
Lithuania, 14, 46–7, 131, 180
Local Agenda 21, xviii–xix, 5, 55–9, 65, 71, 87, 114, 143, 164, 180, 183–4, 186, 189, 190–1, 195–7, 199–202, 228, 238–9
local authorities, 30, 38, 41, 52, 53, 56–8, 86–7, 114, 120, 121–2, 140, 150, 155, 164, 166, 171, 183–4, 188, 193–4, 196, 198–200, 202–3, 227–8, 232–3, 238, 241
Local Government Act (1993) (UK), 58, 200
Local Government International Bureau (LGIB) (UK), 201
Local Government Management Board (LGMB) (UK), 191, 196, 201
local strategies, 5, 45, 56–9, 87, 114, 164, 172, 180, 195, 198, 216, 228
Lucerne, 14, 46, 48
Luton, 196, 201

Maastricht, 4, 43–4, 227–9, 233
Macedonia, Federal Yugoslav Republic of, 14
Major, John, 9, 192

mandates, 24, 102, 107
Manitoba, 55, 59, 90–1
Maori, xvii, 31, 152, 154, 156, 158, 160–1
Martinique, 122
Mediterranean, 118, 122
Mendip, 197
Mexico, 49–50
Milieudefensie, 7, 13, 18–20, 23, 133, 140, 144–6, 233, 249; *see also* Friends of the Earth
Mines Ministers' Conference (Canada), 107
Mining Association of Canada, 107
Ministry for the Environment (MfE) (Netherlands), 4; (New Zealand), 7, 26, 53, 57, 63, 148–9, 157–8, 160–4
Ministry of Agriculture (Netherlands), 133; (Norway), 172
Ministry of Agriculture and Food Economy (Poland), 177
Ministry of Development Cooperation (Norway), 170
Ministry of Economy (Russia), 242–3
Ministry of the Environment and Natural Resources (MoE) (Denmark), 21–2, 40, 52, 56–7, 63, 109–14; (France), 56, 115–24; (Norway), 165, 172; (Portugal, MARN), 241; (Sweden), 41, 182–9
Ministry of Environment (MoE), 28; (Canada), 25; (Denmark), 27, 52, 57, 111–13; (France), 16, 22, 39, 56, 66, 115, 117; (Norway), 166, 168–72; (Russia), 243; (Sweden), 26–7, 41, 189
Ministry of Environmental Protection and Regional Development (MEPRD) (Latvia), 7, 25, 126–32, 249
Ministry of Environmental Protection, Natural Resources and Forestry (Poland), 7, 174, 176–80, 250
Ministry of Finance (Denmark), 112; (Norway), 52, 54, 165, 168, 170, 172; (Sweden), 185
Ministry of Foreign Affairs (Norway), 170, 172
Ministry of Health and Social Welfare (Poland), 177
Ministry of Housing, Spatial Planning and the Environment (VROM) (Netherlands), 7, 38, 57, 128, 133–4, 138–40, 142–3, 233, 249
Ministry of Industry (France), 121
Ministry of Industry and Trade (Poland), 177
Ministry of Labour and Social Policy (Poland), 177
Ministry of Nature Conservation, Agriculture and Fisheries (Netherlands), 143
Ministry of Science and Technological Policy (Russia), 243
Ministry of Spatial Planning and Construction (Poland), 177
Ministry of Transport and Maritime Economy (Poland), 177
Ministry of Works (New Zealand), 162
Minnesota, 55, 205, 216–18
  Minnesota Strategic Plan for Sustainable Development, 218
  Minnesota Sustainable Development Initiative, 217
Misereor, 238
Missouri, 218
Mitterrand, President, 115
Model Communities Programme (UK), 201, 202
Montreal Protocol, 167, 180
Mount Allison University, 59, 91
Muldoon, Prime Minister, 63, 161
Musoma, 207

National Air Issues Coordinating Committee (Canada), 91
National Board of Agriculture (Sweden), 41; *see also* Board of Agriculture
National Commissions on Sustainable Development (European Union), 13, 125
National Committee for International Environmental Issues (Norway), 169
national conservation strategies, 3, 68
National Council for Women (UK), 199

national councils for sustainable development, 13
National Economic and Social Research Institute (Ireland), 239
National Energy Program (Canada), 37
National Environmental Action Plans, 5, 47, 68
National Environmental Action Programme (NEAP), 5, 14, 22, 25–6, 36, 41, 126–7, 129–32; *see also* Environmental Action Programme
National Environmental Education Centre (Poland), 179
National Environmental Goals Project (USA), 7, 27, 214, 220, 251
National Environmental Policy (NEP) (Poland), xvi, xvii, 7, 14, 22, 25–6, 42, 47, 63, 174–81, 250
National Environmental Policy Plan (NEPP), xx, 7, 13, 21, 25–6, 38, 44, 52, 65, 70, 133–44, 146, 224, 228, 249
  NEPP+, 7, 21, 25–6, 133, 135, 138–9
  NEPP2, xx, 7, 16, 21, 25–6, 37–8, 51–2, 58, 70, 133, 135–44, 218
  NEPP3, 53, 70, 133, 136, 139–43
National Environmental Policy Plan for Latvia (NEPPL), 7, 22, 25–6, 39, 48, 60, 126–32
National Environmental Protection Agency (NEPA) (Australia), 87
National Environmental Protection Council (NEPC) (Australia), 76, 87, 89
National Environmental Protection Council Act (Australia), 83, 88
National Environmental Research Institute (NERI) (Denmark), 109
National Forest Policy Statement (Australia), 75, 77
National Institute of Public Health and Environmental Protection (RIVM) (Netherlands), 133, 135, 138, 144
National Landcare Programme (Australia), 77, 85, 89; *see also* Landcare
National Party (New Zealand), 63, 149, 160–1
National Round Table on the Environment and the Economy (NRTEE) (Canada), 7, 13, 32–6, 54, 90, 92, 97, 100–2, 105–6, 108, 247
National Soil Conservation Programme (Australia), 77
National Stakeholders' Assembly (Canada), xvii, 32, 100
National Strategy for Ecologically Sustainable Development (NSESD) (Australia), xvii, xviii, 7, 13, 18, 24, 26, 31, 54, 56, 62, 64, 75, 77–83, 85–9, 247
national sustainable development strategies, 3–6, 9, 56, 59–60, 67–9, 78
National Waterwatch Programme (Australia), 77
Natural Resources and Environmental Protection Cabinet (USA), 217
Natural Resources Canada, 97
Natural Resources Task Force (USA), 40, 212
Nature and Environment Policy (Denmark), xvii, xviii, 7, 9, 22, 27, 40, 51, 57, 63, 110, 113–14, 248
Nature and Forests Agency, 40, 109, 112
Nature Conservancy (Norway), 170
Nature Conservation League (Poland), 179
Nature Protection Act (1991) (Poland), 177
Natuur en Milieu (Netherlands), 133
Netherlands, xv, xvii, 4, 7, 13, 16, 18–21, 23, 25–6, 37–9, 44–5, 51–3, 57–8, 116, 133–6, 140–1, 143–7, 150, 194, 201–2, 218, 221, 228, 233, 249; *see also* National Environmental Policy Plan (NEPP)
Network for Environment and Sustainable Development in Africa (NESDA), xiii
New Brunswick, 55, 90
New Caledonia, 122
New Jersey, 218
New South Wales, 88
New York, 218, 251
New Zealand, xv, xvii, xviii, 7, 13, 16, 18,

24–6, 31, 51, 53, 57, 63–6, 72, 84, 148–64, 250
Newfoundland, 106
non-governmental organizations (NGOs), 27–8, 30–1, 34–5, 38–41, 45, 69, 83, 86, 94, 96, 105–6, 113–14, 117, 123, 128–9, 139–40, 143, 145, 160–1, 164, 166, 169, 171, 175–6, 178–9, 181, 183–5, 189, 193–4, 196, 202, 212, 214, 226–8, 232–5, 238, 241, 243–5, 253
Nordbø, Mrs Eldrid, 170
North American Agreement on Environmental Cooperation, 49–50, 96
North American Free Trade Agreement (NAFTA), xviii, 49–50, 96
North Carolina, 219
North Sea, 167, 172
Northwest Territories, 55, 90
Northwhich, 199
Norway, xv, xvi, 7, 9, 16, 18–20, 22, 24–6, 38–9, 52, 54, 57, 65, 165–7, 169, 170, 173, 250
Nottinghamshire, 198
Nova Scotia, 55, 59, 90–1
NUP (Austria's national environmental plan), 236–7

Office of Policy, Planning and Evaluation (OPPE) (USA), 221, 252
One Billion Trees Programme (Australia), 77
Ontario, 55, 90, 91
Opschoor, Professor J B, 146
Organization for Economic Cooperation and Development (OECD), xiii, 6, 38, 46, 48, 135, 140, 143, 170
Oslo, 172–3, 250
Ottawa, 32, 37, 247
Oxford City Council, 200
Oxford University, 203
ozone, 21, 157, 180, 187, 219, 224

Pakistan, 4, 36
Palmer, Geoffrey, 63, 149
Parliamentary Committee on Planning and Environment (Denmark), 113–14
Parliamentary Office for the Evaluation of Technological Choices (France), 119
ParticipACTION, 102, 105
participation, 3–6, 13, 17, 22, 27, 29–32, 34–6, 39, 41, 47, 49–50, 69–70, 72, 78, 84, 93–4, 100, 104–6, 117, 119, 128–30, 137, 139–40, 145, 161, 171, 178–80, 189, 197–200, 205–6, 211–12, 215, 219–20, 225, 228, 240–1, 245
Partnership for Economic Growth and the Environment (PEGE) (USA), 218
Perth, 88
Plan national pour l'Environnement (PNE), xvi, xviii, 7, 16, 22, 25–6, 38–9, 52, 56, 63, 65–6, 115–18, 121–2, 248
Planning and Compensation Act (1991) (UK), 195
Plymouth, 198
Poland, xv–xvii, 7, 14, 22, 25–6, 42, 46–7, 59, 63–4, 174–6, 178, 180, 250
Polish Ecological Club, 179
Popoff, Frank, 206
Portugal, xx, 45, 228, 241–2
Presidential Honors Program (USA), 206
President's Commission on Environmental Quality (PCEQ) (USA), 206
President's Council on Sustainable Development (PCSD) (USA), xvi, 7, 13, 19, 23–6, 31, 40, 55, 205–7, 208–9, 212–14, 216, 220, 251
Prime Minister's Office (France), 123; (Norway), 52, 168; (Sweden) 170, 186
Prime Minister's Policy Unit (UK), 192
Prince Edward Island, 55, 59, 90–1
Projet de société (Canada), xvi, xvii, 7, 13, 18–19, 24–8, 32–4, 49, 54, 68, 90, 92, 100–2, 104–8, 247

Quebec, 55, 61, 90, 95, 107, 248
Queenstown, 158

Radom Project for Sustainable Communities, 59, 180

Reading, 193, 195
Redclift, M, 56
Rees, William, 20, 107
Regional Environment Centre for Central and Eastern Europe (REC), 14, 47
Renault, 115
Rendezvous (UK), 197
Resource Assessment Commission (RAC) (Australia), 86
Resource Management Act (RMA) (New Zealand), xviii, 7, 13, 16, 18, 24–6, 31, 52–3, 63–6, 72, 84, 148–52, 157–62, 164, 250
Réunion, 122
Rio Chiquito Project, 200
Rio de Janeiro, 13, 20, 33, 41, 97, 100, 102, 104, 166, 170, 182–5, 192, 200, 223, 231
Rocard, Michel, 115
Romania, 46
round tables/round-table meetings, 13, 21, 27–8, 30–2, 34–6, 38, 41, 54–5, 58–9, 64, 69, 83, 90–1, 106, 108, 133, 139–40, 173, 175–6, 178, 194, 196, 199, 200, 203, 212, 217–18, 221, 245; *see also* National Round Table on the Environment and the Economy
Rural and Small Town Research and Studies Programme (Canada), 59, 91
Rural Partnership Programme (Australia), 89
Rushcliffe, 198
Russia, xx, 47, 189, 242–3; *see also* Soviet Union; USSR
Russian Congress on Environmental Protection, 242

Sadler, B, 3, 54, 70
St Lawrence river, 55, 91–2
San Francisco, 4, 212
Saskatchewan, 55, 61, 90
Save the Bush (Australia), 77
Scotland, 196
Scott, Sandy, 108
Sea Monitoring Centre (Latvia), 129
Seattle, 58, 205, 212, 219; *see also* Sustainable Seattle
Shell Canada, 90
Silk, Dana, 36
Silveira, D, 13
Single European Act, 227
Slovakia, 14, 46
SO_2 emissions, 97, 167, 180
Social Democrats, 63, 183
Socialist People's Party (Denmark), 109
Solidarity (Poland), 64, 175
South America, 201
South Australia, 54, 58, 82, 87
South Island (NZ), 157–8
Souviron, Jean Pierre, 123–4
Soviet Union, 22, 46, 63; *see also* Russia; USSR
stakeholder(s), 4–5, 18, 24, 27–8, 30–6, 40, 63, 69, 90–2, 95, 98–100, 102, 105–8, 159–60, 166, 171, 183, 187, 190, 205–6, 212, 221; *see also* National Stakeholders' Assembly
State Committee for Nature Protection (Latvia), 126; *see also* Environment Protection Committee
State Council for Environmental Protection (Poland), 178
State Council for Nature Conservation (Poland), 178
State Environmental Inspection Act (1991) (Poland), 177
State Inspectorate of Environment (Poland), 178
Stevens, J, 20, 193, 251
Stockholm, 9, 43, 250
SustainABILITY, 102, 105
Sustainable Communities Task Force, 31–2, 212, 216
Sustainable Europe, 144, 229, 233, 249
Sustainable Land Management Strategy (SLMS) (New Zealand), 148, 157–8
Sustainable North Carolina Project, 219
Sustainable Seattle, 58, 219; *see also* Seattle
Sustainable Water Management Strategy (SWMS) (New Zealand), 148, 157, 159
Sustainability Round-Table Information Forum (USA), 217

Svalbard (Spitzbergen), 172
Sweden, xv, xvi, xvii, 7, 9, 18–19, 22–7, 41, 44, 57, 63, 65, 182–90, 250–1
  Bill 1993/4:30, 66
  Bill 1993/4:111, xvii, 19, 65, 182, 184, 187, 190, 250
Swedish Association of Local Authorities, 41, 183
Swedish Environmental Protection Agency, 7, 16, 27, 41, 182, 186–7, 251

Tahiti, 122
Tanzania, 198, 200
target groups, 31, 38, 128–30, 135, 137, 139–40, 142, 179, 224, 232
Task Force on Economic Instruments and Disincentives to Sound Environmental Practices (Canada), 34, 93
Tasmania, 14, 76, 78, 89
terms of reference (ToR), 24–7, 83, 93, 102, 115, 129, 139, 161, 171, 185, 189, 194, 203, 221, 244
Theys, Jacques, 117, 248
Toner, G, 32, 36, 50, 61
Town and Country Planning Act (1977) (New Zealand), 53, 149
Towns and Development (T&D) (Netherlands), 202
Treaty of European Union, 43, 227, 229, 231
Tropical Forestry Action Plans, xiii, xix, 68

Ukraine, 47, 180
UN Conference on the Human Environment, 9, 43
UN Conference on Environment and Development (UNCED), xiii, xv–xvi, 3–4, 7, 9, 13–14, 16, 18–21, 24, 27, 37, 41, 44, 47, 58, 61, 65, 67, 78, 83, 86, 90, 92, 96–7, 100, 105, 109, 112, 123, 143–5,165–6, 169–72, 174, 182–3, 185–6, 194–6, 206, 214, 223–4, 227–9, 238, 240, 243–4, 250
UN Conference on Renewable Energy, 106
UN Convention on Biological Diversity, 76, 97
UN Framework Convention on Climate Change, 157, 180
UN General Assembly, 19, 183
United Kingdom (UK), 4, 7, 9, 18, 20, 25–6, 32, 38–9, 43–5, 58, 67, 191–3, 195–203, 228, 251
United Nations Association (UNA), 197, 217
United Nations Commission on Sustainable Development (UNCSD), 88
United Nations Development Programme (UNDP), xiii
United Nations Economic Commission for Europe (UNECE), 46
United Nations Environment Programme (UNEP), 46, 155
Universities Research Group on the Environment, 239
Upton, Simon, 63, 160, 162
Ural Mountains, 46
USA, xv, xvi, xviii, 7, 9, 13, 18–19, 22–6, 31, 40, 47, 49–50, 54–5, 58, 96–7, 205–8, 211–22
USSR, 108; *see also* Russia; Soviet Union

Vale Royal Environment Network (VREN), 199
Vancouver City Plan, 59, 90
Ventspils City, 131
Victoria, 107
Vision 21, 197
Vistula River, 180
voluntary agreements, 38, 53, 136, 139–40, 151, 180, 226
Voluntary Sector Environment Forum (UK), 193

Waitakere, 57, 164
Walesa, President, 177
Wallis and Futuna Islands, 122
Warsaw, 47, 180, 250
Washington DC, 4, 206, 212, 219, 251–2
Western Australia, 82, 87
western Europe, 43, 46

Whitehorse Mining Initiative (WMI), 107
  White House Office on Environmental Policy, 207
  WMI Leadership Council Accord, 107
Winsford, 199
World Bank, xix, 5, 48, 71, 130
World Commission on Environment and Development (WCED), 7, 9, 16, 51, 57, 65, 77, 109, 165–6, 250; *see also* Brundtland Commission
World Conservation Strategy, 18, 39, 55, 77, 91–2
World Conservation Union (IUCN), xix, 4–5, 46, 67, 71, 180
World Health Organization (WHO), 46
World Resources Institute, 209, 251
World Wide Fund for Nature (WWF), 196, 198
  WWF Sweden, 131
Wuppertal Institute for Climate, Environment and Energy, 233, 238

Yeltsin, President Boris, 242
Yukon, 55, 90

For Product Safety Concerns and Information please contact our EU representative GPSR@taylorandfrancis.com Taylor & Francis Verlag GmbH, Kaufingerstraße 24, 80331 München, Germany

Batch number: 10725802

Printed by Printforce, the Netherlands